Russell Library
Middletown, CT 06457

A2140 130348 7

D1200828

Space Shuttle
A Quantum Leap

Space
Shuttle
A Quantum Leap

George Torres

Foreword by
Astronaut Michael Collins

629.441
TOR
AUG 2 6 1988

RUSSELL LIBRARY
123 BROAD STREET
MIDDLETOWN, CT 06457

PRESIDIO PRESS ★ NOVATO, CALIFORNIA

Published by Presidio Press, 31 Pamaron Way, Novato, CA 94947

Except in the United States of America, this book is sold subject to the condition that it should not, by way of trade or otherwise, be lent, resold, hired out, or otherwise circulated without the publisher's prior consent in any form of binding or cover other than that in which it is published, and without a similar condition, including this condition, being imposed on the subsequent purchaser.

Library of Congress Cataloging-in-Publication Data

Torres, George J., 1954-
   The space shuttle, a quantum leap.

   1. Space shuttles. I. Title.
TL795.5.T67 1986   629.44'1   85-30175
ISBN 0-89141-253-0

Printed in Japan by Dai Nippon

ACKNOWLEDGMENTS

What are they doing up there? That oft-repeated question led me to write The *Space Shuttle: A Quantum Leap.* Of course no book on a subject as broad as the space program can be undertaken without the help of many people. Special thanks go to the companies and organizations that graciously supplied me with the wonderful photographs that illustrate the book: Rockwell International, The JFK Library (Boston), Hughes Aircraft, The Jet Propulsion Laboratory, and the White House.

A heartfelt thanks goes out to all the friends and colleagues who read early versions of the manuscript and encouraged me to continue—I couldn't have done it without you.

Contents

Dedication

This book is dedicated to the memory of the courageous *Challenger* crew who lost their lives on January 28, 1986:

Commander:	Francis (Dick) Scobee
Pilot:	Michael Smith
Mission Specialist:	Judith Resnik
Mission Specialist:	Ellison Onizuka
Mission Specialist:	Ronald McNair
Payload Specialist:	Gregory Jarvis
Payload Specialist:	Sharon (Christa) McAuliffe

The Space Shuttle: A Quantum Leap was set to go to print the week of the *Challenger's* accident. It is an optimistically written account of the potential of the U.S. Space Program. The loss of *Challenger* may affect the timetable of upcoming space activities, but I am sure that the crew did not die in vain and the space shuttle program will recover from this tragedy to reach new heights.

"We've grown used to wonders in this century; it's hard to dazzle us. But for twenty-five years the United States program has been doing just that. We've grown used to the idea of space and perhaps we forget that we've only just begun—we're still pioneers. They, the members of the *Challenger* crew, were pioneers.

We'll continue our quest in space. There will be more shuttle flights and more shuttle crews and, yes, more volunteers, more civilians, more teachers in space. Nothing ends here. Our hopes and our journeys continue."

President Reagan's Address to the Nation
January 28, 1986

Challenger's last crew: (*clockwise from top left*) Ellison Onizuka, Sharon (Christa) McAuliffe, Gregory Jarvis, Judith Resnick, Ronald McNair, Francis (Dick) Scobee, Michael Smith.

FOREWORD

When I became an astronaut in 1964, I had only a vague understanding of where Project Mercury would lead us. To a lunar landing, certainly, because President Kennedy had already made his stirring speech to the Congress on the subject. But toward a huge airplane-like vehicle called the *shuttle?* Hardly. In this fast-paced account, George Torres explains how this came to pass.

Throughout the book, the shuttle plays center stage but the genesis of this hybrid airplane-spacecraft is put into careful perspective by a strong supporting cast. For example, he shows how Gemini bridged the gap between Mercury and Apollo, and how each of these programs employed increasingly advanced technology which permitted the sophistication of the systems we find on the shuttle today.

The author's enthusiasm for the shuttle and the future of manned space flight permeates this book; but this enthusiasm hasn't led him to neglect the unmanned scientific probes. For example, he vividly describes the fascinating Ulysses mission that in 1990 and 1991 will look down on the sun's poles and all the rest of the solar system—a view different from any we have seen to date.

Just as President Kennedy defined a trip to the moon, President Reagan has put his personal stamp on the space station, which will create a permanent human presence in earth orbit a decade from now. The Europeans and Japanese have already made a financial commitment to this idea and will contribute laboratory modules and other components. In all likelihood the crew of six or eight will also be international. Beyond the space station it is difficult to define the next logical step. But one thing is clear, this book gives the reader a firm foundation upon which to judge the progress we have made so far and therefore to assess future endeavors we might undertake. It also touches on some of the military aspects of space.

A Quantum Leap will also find a place as a handy and concise reference volume. It lists the American manned flights—six Mercury, ten Gemini, eleven Apollo, and three Skylab—by date, crew, duration, and distance travelled. A similar shuttle chronology is included but, of course, that list will continue to grow for years. Planetary missions of the greatest importance are also listed, including what may be the most historic of all flights, that of *Pioneer 10,* the first man-made object to escape our solar system. An end-to-end shuttle mission profile, a directory of NASA centers, and the companies across the nation that built the shuttle complete this copiously illustrated, well-rounded volume.

From Sputnik to celestial cities, George Torres has outlined the important milestones along the path, and emphasized the pivotal role that the shuttle has, and will, play in our future.

Michael Collins

CHAPTER 1

A QUANTUM LEAP INTO THE FUTURE

"EARTH IS THE CRADLE OF MANKIND, BUT MAN CANNOT LIVE IN THE CRADLE FOREVER."

KONSTANTIN TSIOKOVSKY (1857–1935) (EARLY RUSSIAN ROCKETRY THEORIST)

The past three decades have been an extraordinary period of human growth in all fields. Incredible achievements. The computer age has evolved from infancy through four generations and we're on the verge of a fifth; in aviation, we have gone from the DC-3 through the 747 and the Concorde; and in medicine, heart transplants are today a common occurrence. Yet, space exploration is probably the single endeavor that has yielded the most profound advances. In just three short decades we have learned more about the universe, our own solar system, and indeed our own planet, than in all previous recorded history.

We have launched hundreds of satellites, sent mechanical visitors to other planets, and have seen men walk on the moon. Great strides have indeed been made, and we now stand on the threshold of a new era in space exploration that will make space travel seem routine. The space shuttle is the first step in making that era a reality.

Because the space shuttle can do a wide variety of tasks impossible with earlier launch vehicles, it will be the space program's workhorse for the 1980s, 1990s, and into the next century. The shuttle's unique capabilities open new windows of opportunity in space exploration. The launch of larger, more capable satellites, on-orbit satellite repair and servicing, breakthroughs in zero-gravity materials processing, and construction of space stations are all within reach. To that end an initial fleet of four shuttle orbiters has been built that will enable the frequency of shuttle flights to increase to twenty-four missions per year by the late 1980s. Contemplated additions to the fleet would allow thirty to forty missions yearly in the 1990s.

The space shuttle's unique capabilities will fundamentally change our ability to utilize the space environment, heretofore limited by launch-system capabilities. Earlier launch vehicles had payload capacities a fraction of the shuttle's. In addition, they could only *deliver* cargo to orbit; the shuttle can *return* cargo to earth. With its spacious cargo bay the shuttle is able to transport a wide variety of satellites and experiments into

Orbiter as photographed by the SPAS satellite, just deployed by the shuttle's robot arm.

1

orbit, many of which could not be launched before due to the size, shape, or weight constraints imposed by earlier launch vehicles.

Today's objective, however, is not simply to launch cargo into orbit, but to actually have crews working in space. And perhaps the space shuttle's greatest advantage is that it can transport a crew of up to eight persons. Astronauts, either inside the shuttle or outside in space suits, can perform tasks that would be impractical or impossible with any other launch vehicle.

The shuttle's large cargo capacity and crew accommodations are allowing us to rethink our entire approach to satellite design and the way we operate in space. Unique shuttle capabilities include:

- VERSATILITY—The shuttle's large 60 foot by 15 foot payload bay enables it to deliver 65,000 pounds of cargo to orbit. Several satellites can be carried on each mission, thereby reducing the cost to satellite owners since they share the launch costs. Other payloads such as telescopes, sensors, planetary probes, or any combination of experiments can share a mission. In addition, the shuttle can carry Spacelab: a scientific laboratory that transforms the shuttle into a mini space station.

- SATELLITE RETRIEVAL, REPAIR, AND RESUPPLY—The shuttle-based repair of the *Solar Maximum* satellite in April 1984 began a new era for satellite designers. They can now build their satellites on a modular basis with repair and upgrading in mind. Other spacecraft, such as the Space Telescope, have already been designed for routine on-orbit servicing to extend their useful lives. Scheduled for launch in 1986, the Space Telescope is designed for servicing and resupply on a

The shuttle's two-level crew cabin can accommodate up to eight astronauts.

regular basis, or it can have its components replaced with advanced units as they are developed.

- REPEAT MISSIONS—In contrast to satellites that are deployed by the shuttle, many shuttle experiments remain in the cargo bay and are returned to earth with the orbiter. New satellite components, such as sensors or antennas, can be tested aboard the shuttle and returned to earth for modification and reuse in space. This will accelerate the development of satellite technology by obviating the building of unique and expensive satellite carriers each time a new component design requires testing.

- ON-ORBIT ASSEMBLY—Shuttle crews will make it possible to assemble enormous satellites and even space stations in orbit. Components will be carried up in multiple shuttle missions and assembled

Above: Astronaut James Van Hofton replaces Solar Max's main electronics box, while George Nelson hovers along the bottom of the payload bay.

in space by the astronauts, much like building with giant erector sets.

Space Shuttle Mission Profile

We have all seen the spectacular television coverage of a shuttle launch and landing, but a typical shuttle mission goes through many phases that television cannot cover. Large parts of the following section are excerpted from the actual space-to-ground tapes between mission control and astronauts John Young and Bob Crippen during the first shuttle mission.

As each shuttle achievement is reached, the mission seems almost routine. Nothing could be further from the truth.

It all begins in the enormous fifty-two-story Vehicle Assembly Building where the orbiter is attached to the external fuel tank and twin solid rocket boosters. Once this amazing vehicle is thoroughly checked it begins a 3½-mile journey to the launch pad at the ground-crunching pace of approximately 1 mile per hour.

On the launch pad, extensive checkout and a series of practice countdowns are performed over a two week period. Six hours before launch the large external tank is loaded with over half a million gallons of liquid hydrogen and oxygen. The shuttle is now ready for launch.

By sunrise on launch day, the crew already has been strapped into the orbiter for two hours, checking and rechecking all systems. The entire shuttle team is now ready—mission control in Houston, ground crews at landing sites around the world, even the contractors around the nation who built the shuttle—and stand ready to provide technical support in case of problems.

Launch Control (LC): T minus 1 minute 10 seconds and counting. Liquid hydrogen tank is at flight pressure. T minus 45 seconds and counting. T minus 40 seconds and counting. T minus 20 seconds and counting. T minus 15, 14, 13, T minus 10, 9, 8, 7, 6, 5, 4, we have gone for main engine start . . . We have liftoff of America's first space shuttle . . . and the shuttle has cleared the tower.

Another mission is on as seven million pounds of thrust hurl the shuttle skyward. The shuttle clears the launch tower, turns out over the Atlantic, and quickly accelerates. Mach 1, Mach 2, Mach 3. At the two minute mark the shuttle is travelling over 3,000 miles per hour and has reached an altitude of 28 miles.

Capsule Communicator (CAPCOM) at Mission Control in Houston: *Columbia,* Houston, you're go at throttle up.

Columbia: Roger, go at throttle up.

Public Affairs Officer (PAO): 1 minute 45 seconds, coming up on go—no go.

CAPCOM: *Columbia,* you're negative seats.

PAO: That callup says that *Columbia*'s altitude is too high for ejection seat use.

CAPCOM: *Columbia,* you are go for SRB [solid rocket boosters] separation.

PAO: 2 minutes 4 seconds, standby for SRB sep confirmation.

The twin solid rocket boosters have completed their job and are jettisoned for recovery by waiting ships.

PAO: Mark 2 minutes 30 seconds, onboard guidance system converging as programmed. *Columbia*

Left: Orbiter just prior to mating with external tank and solid rocket boosters.

is now steering for its precise window in space for main engine cutoff. Mark 2 minutes 40 seconds. *Columbia* now 39 nautical miles in altitude 42 nautical miles downrange. Mark 2 minutes 50 seconds.

CAPCOM: *Columbia,* you're looking a little hot. All your calls will be a little early.

Above: The 1 mph ground-crunching trip to the launch pad.
Right: Accelerating over the Atlantic.

SRB separation: notice the small thrusters used to push the solid rocket boosters away from the orbiter.

Columbia: Okay.

PAO: *Columbia* now has two engine Rota capability . . . looks good here. [Abort capability to Rota, Spain contingency landing site now possible with just two engines]

PAO: Mark 3 minutes. Young and Crippen really moving out now, velocity now reading 6,200 feet per second. Mark 3 minutes 15 seconds. *Columbia* now 51 nautical miles in altitude, 66 nautical miles downrange, velocity now reading 6,500 feet per second. Mark 3 minutes 30 seconds, *Columbia* now 55 nautical miles altitude, 78 nautical miles downrange.

Having dropped its heavy SRBs the orbiter is

really accelerating now—Mach 5, 6, 10, 12, 15.

PAO: *Columbia* given a green to continue. Mark 3 minutes 55 seconds standing by for a press to MECO [Main Engine Cutoff].

CAPCOM: Ready for press to MECO.

Columbia: Roger, press to MECO.

PAO: Mark 4 minutes 8 seconds.

CAPCOM: *Columbia,* stand by for negative return. Mark negative return.

PAO: Mark 4 minutes 25 seconds with that callup from CAPCOM Brandenstein. *Columbia* now committed to space travel. Young and Crippen can no longer turn around and return to launch site.

Columbia: What a view! What a view!

CAPCOM: Glad you are enjoying it.

PAO: Mark 5 minutes 15 seconds. *Columbia* now 75 nautical miles altitude, 202 nautical miles downrange.

CAPCOM: *Columbia,* you are single engine press to MECO.

PAO: Mark 7 minutes 20 seconds. That report says that Young and Crippen can achieve orbit insertion even if two engines go out. Mark 7 minutes 30 seconds. *Columbia* 67 nautical miles altitude, 485 nautical miles downrange. G-force is building for

Young and Crippen, now. Up to 3 g's. Mark 7 minutes 45 seconds. *Columbia's* main engines slowly need throttle back now, should be throttled at 65 percent, that is 6 seconds before main engine cutoff.

OMS (orbital maneuvering system) engines firing; this finalizes shuttle orbit.

After traveling for almost eight minutes the shuttle is moving over 17,000 miles per hour at an altitude of over 70 miles. Here the orbiter shuts down its main engines, separates from the external tank, and prepares to circularize its orbit.

The astronauts are now riding in a quiet spaceship. No more boosters. The main engines are silent. The crew is tempted to relax. Almost. It's time to use the orbital maneuvering system (OMS) to go into orbit. The OMS's two small engines provide the shuttle's maneuvering capability once in orbit.

In orbit the crew quickly opens the payload-bay doors so that the door radiators can be used to dissipate the internal heat buildup from the orbiter's electronic systems. The doors must remain open throughout the mission. If they cannot be opened the orbiter must return to earth or it will overheat.

Columbia: You're missing one fantastic sight. Here comes the right door and boy that is really beautiful out there.

CAPCOM: We appreciate the great view updates.

Columbia: Roger that.

CAPCOM: Right door now open.

Columbia: All the latches work just fine and the door looks like she's doing her thing.

With the doors open the crew is now ready to begin carrying out their mission objectives. A typical shuttle mission lasts approximately seven days during which time the orbiter will travel almost 3 million miles as it orbits the earth between ninety and one-hundred-ten times.

However, each mission is anything but typical. Any single mission might carry three or four satellites in the cargo bay, each worth up to $100 million. Or the objective might be to retrieve and repair a satellite. There will probably be experiments in the crew compartment to produce unique materials only possible in zero gravity. Another mission might carry Spacelab, the European facility that turns the shuttle into a mini space station. Still others will carry planetary probes for deployment as they begin their reconnaissance of the solar system, or large space telescopes expected to see to the edges of the universe once the shuttle stations them above our murky atmosphere.

As the mission nears its end the doors are closed and the orbiter is prepared for atmospheric reentry. The orbiter must first be slowed by several

hundred miles per hour by placing it in a tail-first position and firing the two orbital maneuvering system engines.

Once the orbiter slows sufficiently it is turned back to a nose-forward position. Reentry begins at about 400,000 feet (80 miles), where the atmosphere begins to quickly reduce the speed of the orbiter. The orbiter is now traveling just below Mach 25 (17,500 miles per hour) and is over 5,000 miles from its landing site. It soon resembles a flaming meteor, hurtling through the atmosphere.

PAO: This is Mission Control. Guam here has lost its signal. *Columbia* is 1 minute 32 seconds away from entering the earth's atmosphere. We are showing 34 minutes 21 seconds to touchdown at Edwards Air Force Base. We will be out of communication with *Columbia* for approximately 21 minutes. No tracking stations before the West Coast. And there is a period of about 16 minutes of aerodynamic reentry heating that communications are impossible during this entry. *Columbia* in good shape and the crew in good shape for this entry.

Left: The glider returns.

Above: Ground crews first purge all remaining fuel as they begin preparation for the next flight.

Having survived the flaming reentry the orbiter operates as a glider to approach the landing strip. It uses its wing elevons to make a series of energy dissipating turns that reduce its speed to approximately 210 miles per hour for touchdown.

Chase Plane
One: They're down [landing gear], pick up your feet. Five, four, three, two, one, touchdown.

Welcome home, *Columbia*. Beautiful, beautiful.

Columbia: Do I have to take it up to the hangar, Joe?

CAPCOM: We're going to dust it off first.

Columbia: This is the world's greatest flying machine, I'll tell you that. It worked super.

Once the Orbiter comes to a stop, ground crews quickly begin preparing it for its next mission, and the next, and the next.

CHAPTER 2

THE FIRST STEPS

"FUTURE GENERATIONS WILL SAY THAT THE REAL SIGNIFICANCE IN OUR SPACE PROGRAM LAY IN THE FACT THAT IT TOOK THE 'LID OFF' THE LIMITATIONS POSED BY THE FINITE SIZE AND FINITE RESOURCES OF THE PLANET EARTH."

WERNHER VON BRAUN

Before exploring exactly how the space shuttle will take us a *quantum leap into the future,* we must first revisit the exciting beginnings of the space program. The shuttle would have been an impossibility had it not been for breakthroughs made in earlier space projects.

The space age began as an outgrowth of the mid-1950s cold war between the United States and the Soviet Union. Driven by the desire for military and technical leadership, the two countries raced to put a satellite into orbit. Both sides feared that a rocket capable of putting a satellite into orbit could also be used to deliver a nuclear weapon.

The original seven Mercury astronauts:
Clockwise from top left, Alan Shepard, Gus Grissom, Gordon Cooper, Scott Carpenter, John Glenn, Deke Slayton, Wally Schirra.

After all, an airplane carried the first atomic weapons into enemy territory. Rockets capable of reaching space offered the spectre of nuclear warfare by remote control.

Many also recognized the great potential of space for peaceful purposes. As early as 1945, Arthur C. Clarke, the British scientist and writer, accurately described in his landmark article in *Wireless World* how satellites could provide continuous worldwide radio coverage if placed in an orbit that matched the earth's rotation. As with most new technology, the reality took a few years to catch up to the idea.

The Russians were first to announce their intention of putting a satellite into orbit. On April 15, 1955, the Soviet press reported that a "Commission on Interplanetary Communications has been established to develop satellites for meteorological purposes." This announcement was largely ignored by the world press. It wasn't until July 29, 1955, when the White House announced that the United States would orbit a satellite, that the "space race" stirred attention. This announcement forced the world to sit up and take notice. When the Soviets repeated their earlier announcement the next day, the world responded with a collective "Sure you will." Who could believe it? After all, the "Russian Bear" was known to be slow and technologically backward.

Left: Vanguard 1—AKA "Kaputnik."
Right: Alan Shepard becomes the first American to venture into space, May 5, 1961.

This condescending attitude may have been just the incentive the Russians needed. On October 4, 1957, *Sputnik,* the first artificial satellite to orbit the earth, inaugurated the space age. Although Sputnik was tiny—180 pounds—compared to current satellites, the Russian achievement shocked the world. For the United States it was a technological slap in the face.

The American people were outraged and frightened. *Sputnik* could be seen glinting in the sky as it passed overhead like clockwork every ninety minutes. Then less than a month later the Russians launched *Sputnik-2,* a much larger 1,200-pound spacecraft that carried a dog as its passenger.

Meanwhile America's rocket development was dragging along. The first American attempt to launch a satellite took place on December 6, 1957, and resulted in humiliation. The Vanguard rocket exploded 4 feet off the launch pad. Newspaper headlines screamed "KAPUTNIK!"

Washington frantically searched for an alternative and found one at the Army Ballistic Missile Agency in Huntsville, Alabama. There Wernher von Braun—the German missile genius who developed the V2 rockets during World War II—had assembled a team and was in the process of expanding the basic V2 into a rocket that could reach earth orbit. He accepted the challenge of launching an American satellite and redoubled his team's efforts.

On January 31, 1958, after months of skepticism from the press and the White House, America's first satellite, *Explorer I,* was successfully launched aboard von Braun's Jupiter C rocket. *Explorer I* also scored a scientific success as it discovered the Van Allen radiation belt, the belt of intense ionizing radiation that surrounds the earth in the outer atmosphere. Physicist James Van Allen had theorized that such a belt existed; *Explorer I* confirmed it. Two other Explorer satellites and a Vanguard were also successfully launched in 1958 as the space race began picking up momentum.

NASA and the Mercury Program

The next step in the space race was to put a man into space. To further that goal President Eisenhower established the National Aeronautics and Space Administration (NASA) on July 29, 1958. NASA officially became an agency on October 1, 1958, and was given a broad mandate to create a comprehensive program of aeronautical and space goals that would establish U.S. leadership.

One of NASA's first undertakings was the Mercury program. Mercury's "modest" goals included placing an astronaut into orbital flight around the earth, investigating his adaptation to a zero-gravity environment, and recovering the capsule and astronaut safely. While this sounds simple today, when the Mercury program began there were countless unknowns. What would zero-gravity do to a person's blood circulation or breathing? No one knew.

Over the next three years, the challenge of developing new technologies to conquer the unknown frontier of space engulfed the scientific community and the American people. Hardware had to be designed and constructed that was capable of protecting a man in the harsh, airless environment of space and returning him safely through the burning atmospheric reentry to a splashdown in the ocean. The space race had entered its second chapter.

As progress was made, the idea of sending a man to the moon began to form in the minds of many. Since ancient times humanity had looked at the moon and yearned. In fact, it was said that Russian president Nikita Khrushchev wanted to send a man to Mars! However, it was painfully obvious that it would take a tremendous effort just to get a man into orbit. Mars and the moon would have to wait.

The First Man in Space

Once again the Russians scored first, this time by successfully putting the first man into orbit, Yuri Gagarin. Gagarin's historic flight lasted all of one hour and forty-eight minutes as he made one complete orbit of the earth on April 12, 1961. For most Americans, Gagarin's single orbit of the earth was another demonstration of Soviet technological advancement.

Yet, at the time of Gagarin's flight, NASA knew it would be attempting its own manned flight in less than a month. Prior NASA launches of unmanned Mercury spacecraft, and others that carried chimpanzees, gave the agency the confidence it needed to weather the storm of criticism that followed Gagarin's flight.

On May 5, 1961, NASA's confidence was con-firmed as Alan Shepard became the first American to enter space. While Shepard's flight was only a fifteen-minute partial orbit, this was according to plan. The Redstone rocket that boosted Shepard into space was only capable of putting the Mercury capsule into a short arcing trajectory. The flight was intended to verify the Mercury capsule for upcoming orbital flights using the much larger Atlas rocket.

Shepard's pioneering effort did more than put an American in space and make him a national hero. Shepard's success prompted President Kennedy to go before a joint session of Congress to pledge his commitment to carry out one of mankind's oldest dreams. On May 25, 1961, less than three weeks after Shepard's flight, the president delivered these immortal words:

Now it is time to take longer strides—time for a great new American enterprise—time for this nation to take a clearly leading role in space achievement, which in many ways may hold the key to our future on earth.

I believe that this nation should commit itself to achieving the goal before this decade is out, of landing a man on the moon and

President Kennedy awards Alan Shepard the Distinguished Service Medal.

"I believe we should go to the moon."

returning him safely to earth. No single space project will be more exciting or more impressive to mankind, or more important for long range exploration of space, and none will be so difficult or expensive to accomplish.

Gemini Steps Up the Pace

At the time of President Kennedy's announcement, Alan Shepard's fifteen minutes of spaceflight was the only experience NASA had on which to base its decisions on sending a man to the moon and returning him safely to earth. It would take more than a week to complete the half-million mile round-trip to the moon. Many procedures, such as spacecraft rendezvous and docking, and extravehicular activity (spacewalks), had to be perfected. It was also necessary to gain a greater understanding of the effects of long-duration spaceflight on the astronauts. The Gemini program was initiated specifically to answer those questions.

While Gemini spacecraft were being built, five other Mercury missions were conducted. Shepard's suborbital flight was followed two months later by a similar fifteen-minute mission flown by Gus Grissom. Then on February 20, 1962, John Glenn became the first American to orbit the earth. Glenn's three orbits made him perhaps the most famous of the early astronauts and gave the American people the confidence that their country could, indeed, send a man to the moon. Three more flights, each of longer duration than Glenn's, would complete the Mercury program by May 1963.

Almost two years later the Gemini flights began. Consisting of ten manned flights conducted in rapid succession between March 1965 and November 1966, the Gemini program was designed to bridge the enormous gap between the relatively simple Mercury earth-orbital program and an infinitely more complex mission to the moon. Gemini missions would investigate many of the critical procedures required in a voyage to the moon.

Whereas the Mercury capsule had carried only one man and had small jets to control the altitude of the spacecraft, the larger Gemini spacecraft would carry two men and have an onboard propulsion system for orbital maneuvers. Gemini spacecraft were also equipped with a guidance and navigation system and a rendezvous radar, thus enabling the astronauts to try out various rendezvous and docking techniques essential for lunar missions. During *Gemini 6* through *Gemini 12,* the complex task of rendezvous and docking two vehicles in space was largely perfected.

The Gemini program provided NASA with invaluable experience in extravehicular activity (EVA). Ed White became the first American to walk in space during *Gemini 4,* and the word *spacewalk* was added to the dictionary. EVAs also took place during *Gemini 9* through *Gemini 12.*

The Gemini program achieved all of its objectives and gave the United States nearly two-

The Atlas launch that made John Glenn the first American to orbit the earth, February 20, 1962.

JFK and John Glenn examine the *Friendship 7* Mercury capsule that carried Glenn on his historic mission.

Mercury Missions

Mission	Date	Astronauts	Mission Duration Hrs:Min:Sec	Earth Orbits	Miles Traveled	Launch Vehicle
Mercury 3*	May 5, 1961	Alan Shepard	00:15:22	Suborbital	625	Redstone
Mercury 4	Jul 21, 1961	Virgil "Gus" Grissom	00:15:37	Suborbital	629	Redstone
Mercury 6*	Feb 20, 1962	John Glenn	04:55:23	3	80,787	Atlas
Mercury 7	May 24, 1962	Scott Carpenter	04:56:05	3	81,146	Atlas
Mercury 8	Oct 3, 1962	Walter Schirra	09:13:11	6	153,564	Atlas
Mercury 9	May 15–16 1963	Gordon Cooper	34:19:49	22	582,181	Atlas

*Two Mercury-Redstone and two Mercury-Atlas flights preceded the manned missions. Each launch vehicle type carried one unmanned Mercury capsule and one with a chimpanzee aboard.

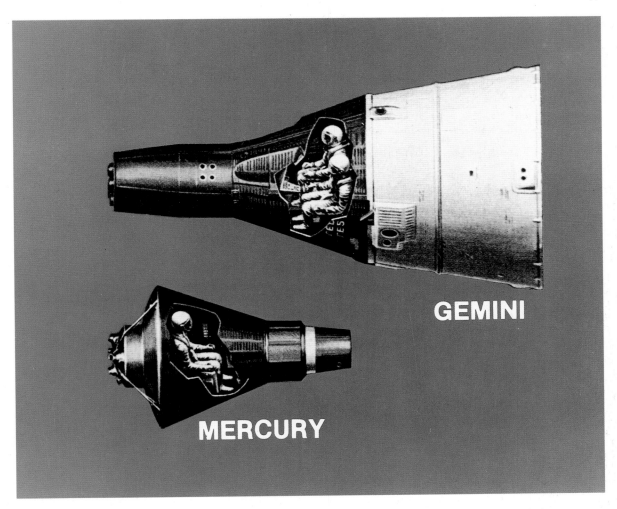

GEMINI

MERCURY

Left: Ed White became the first American to walk in space during this 21-minute spacewalk, June 3, 1965. In his right hand he is holding a self-maneuvering propulsion unit.

Above: Size comparison of the Gemini and Mercury capsules.

thousand hours of man-in-space experience. It prepared the mission control team for the Apollo program, and finally, the Gemini program bolstered the nation's confidence in its ability to voyage to the moon.

Scouting the Route

Even before President Kennedy committed the United States to the Apollo program, NASA had planned to send automated spacecraft to the moon. In 1959 and 1960, three *unmanned* programs—Ranger, Surveyor, and Lunar Orbiter —were established to investigate the moon. Once landing a man on the moon became the an-

Gemini Missions

Mission	Launch Date	Astronauts	Mission Duration Days:Hrs:Min	Earth Orbits	Miles Traveled	Launch Vehicle
Gemini 3*	Mar 23, 1965	Virgil "Gus" Grissom John Young	00:04:53	3	79,823	Titan II
Gemini 4	Jun 3–7, 1965	James McDivitt Edward White	04:01:56	62	1,606,148	Titan II
Gemini 5	Aug 21–29, 1965	Gordon Cooper Charles Conrad	07:22:55	120	3,330,634	Titan II
Gemini 7**	Dec 4–18, 1965	Frank Borman James Lovell	11:18:35	206	5,704,785	Titan II
Gemini 6	Dec 15–16, 1965	Walter Schirra Thomas Stafford	01:01:51	16	448,807	Titan II
Gemini 8	Mar 16, 1966	Neil Armstrong David Scott	00:10:41	7	181,049	Titan II
Gemini 9	Jun 3–6, 1966	Thomas Stafford Eugene Cernan	03:00:21	44	1,252,859	Titan II
Gemini 10	Jul 18–21 1966	John Young Michael Collins	02:22:47	43	1,220,670	Titan II
Gemini 11	Sep 12–15 1966	Charles Conrad Richard Gordon	02:23:17	44	1,229,810	Titan II
Gemini 12	Nov 11–15 1966	James Lovell Edwin "Buzz" Aldrin	03:22:35	59	1,596,469	Titan II

*The first two Gemini missions were unmanned
**Gemini 7 was planned as a long duration mission and was launched first to serve as a rendezvous target for Gemini 6

nounced goal, these programs took on much greater significance. They would scout the route Apollo would follow years later. Ranger would provide a close-up look at the moon's surface; the Lunar Orbiter would map the planet for landing sites; and Surveyor would land on the moon to determine its composition and firmness.

The Ranger program consisted of kamikaze-type missions designed to photograph the moon while enroute, and at close range, prior to crashing into the surface. The difficulties that awaited Apollo became glaringly apparent during the Ranger Program—its first *six* launches resulted in failure. Upper stages wouldn't fire (or misfired), computers failed, and even the cameras refused to operate. Moreover, the Rangers were tiny by Apollo standards so theoretically their relatively small rockets *should* have been easier to control.

Yet many lessons were learned during those failures, and Ranger's final three attempts between July 1964 and March 1965 were spectacular successes. These three Ranger spacecraft returned over 17,000 photos of the moon's surface, revealing details heretofore unseen on the earth. Included were excellent photos of the Sea of Tranquility, which would become the first Apollo landing site.

The Lunar Orbiter program, the second of the automated lunar explorers, began its work on August 10, 1966. Its five missions, which continued through August 1967, provided excellent coverage of all twenty potential landing sites for Apollo and confirmed the Sea of Tranquility as the first target. In addition, they provided more detailed photographs of the back side of the moon than any earth-based telescope had ever provided of the front side.

The Surveyor program, the third member of the team, was designed to measure the firmness of the moon's surface and confirm its ability to withstand a landing similar to that planned for the much larger Apollo lunar module. Five of the seven Surveyor missions between June 1966, and January 1968, were successful in providing detailed information on specific landing sites. Although the Surveyor spacecraft were much smaller than those planned for Apollo, the landings showed that the lunar surface could support a spacecraft. Prior to the Surveyor landings, no one knew just how deep the moon's top layer of dust was or if a spacecraft would sink into it and disappear.

With the completion of the Ranger, Lunar Orbiter, and Surveyor programs, the unmanned scouting of the route was complete. Confidence was growing. Few surprises were expected to confront the first men on the moon.

Left: A Ranger spacecraft — Kamikaze missions.
Right: Surveyor VII lunar landing craft.

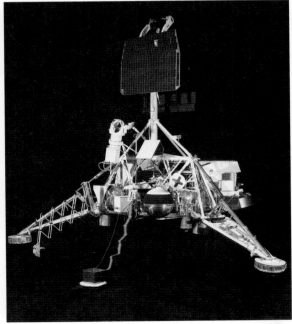

CHAPTER 3

APOLLO: MANKIND'S GREATEST ADVENTURE

"NOW IS THE TIME TO TAKE LONGER STRIDES . . . TIME FOR A GREAT NEW AMERICAN ENTERPRISE . . . TIME FOR THIS NATION TO TAKE A CLEARLY LEADING ROLE IN SPACE ACHIEVEMENT, WHICH IN MANY WAYS MAY HOLD THE KEY TO OUR FUTURE ON EARTH."

PRESIDENT JOHN F. KENNEDY ADDRESSING A JOINT SESSION OF CONGRESS, MAY 25, 1961

Could it be done? What would it take to land men on the moon? The Mercury and Gemini programs were daring and dangerous—with Apollo the risks would be multiplied many times. The trip to the moon would be unforgiving. In case of an emergency, Mercury and Gemini astronauts could return to earth in a matter of minutes—it would take three *days* for an Apollo crew to return from the moon.

The Apollo challenge was to create a spacecraft large and complex enough to supply all the needs

An earthrise as seen by the *Apollo 8* crew, December 1968.

of three astronauts for two weeks. This home-away-from-home had to contain all the life-sustaining elements—food, air, and shelter—and it had to be *totally* dependable. A mishap during a lunar mission could spell disaster. The answer to this enormous challenge: the Apollo command and service modules.

An equally challenging question was how to get this three-man crew to the moon, 239,000 miles away. To travel the enormous distance between the earth and the moon, the Saturn V—the largest, most powerful launch vehicle ever designed—had to be built.

A comparison of Gemini and Apollo launch requirements reveals the daunting task of designing the Saturn V. The Titan II rockets that boosted the Gemini spacecraft into *earth orbit* stood 103 feet tall and had to carry 3 tons of men and equipment at 17,500 miles per hour to break out of the earth's atmosphere. Apollo missions required that 45 tons be carried to *lunar orbit*.

The trip to the moon was essentially a two-step maneuver. First, the Saturn V would hurl 150 tons of Apollo equipment into earth orbit. Then it would fire another stage to boost 45 of those tons to a speed of over 24,000 miles per hour in order to escape the earth's gravitational pull and send

Above: Apollo's "home away from home," a command module in the foreground and one mated to the service module on the overhead hoist, at the North American Aviation assembly plant, Downey, California.
Below: A lunar module undergoing checkout.

the spacecraft on its way to lunar orbit. The 363-foot tall, three-stage Saturn V was powered by five mammoth first-stage engines that generated a total of 7,500,000 pounds of thrust. The Saturn V's second- and third-stage engines provided an additional 1,125,000 and 225,000 pounds of thrust respectively. By comparison, the two-stage Titan II needed only 430,000 and 100,000 pounds of thrust to power its stages.

The third major component of the Apollo program was the lunar module. While the Saturn V boosted the command module and its three-man crew into lunar orbit, the lunar module—a spider-like landing vehicle—would actually carry two of the astronauts down to the lunar surface and then return them to the orbiting command module after they finished their surface exploration. Its single boost engine had no margin for error. If it failed, the astronauts could be stranded on the moon.

Apollo Testing

The developmental phase of the Apollo program proceeded smoothly during the early 1960s, and it continued to pick up steam even as Gemini missions were being flown. Successful engine tests were conducted, and the Apollo command and service modules were steadily passing all the tests. In 1966 two unmanned Apollo spacecraft were successfully launched into earth orbit to test the reentry heat shield and the operation of other subsystems.

By January 1967, Apollo was almost ready for its first manned test flight when tragedy struck. During a routine flight-simulation test on the launch pad, a fire broke out in the command module. The fire apparently was sparked by an electrical short circuit that ignited the oxygen-rich atmosphere. The three astronauts aboard, Gus Grissom, Ed White, and Roger Chaffee, died of asphyxiation.

Left: The 33-story Saturn V on the launch pad.

Right: Apollo 7, testing the new system.

The Apollo program came to a screeching halt and many called for an end to the impossible dream. President Kennedy was no longer around to support the program. The Vietnam war was getting out of control. And many Americans were becoming more hesitant to support a space program at a time when there were so many problems on earth. Yet, President Kennedy's words could still clearly be heard.

It would be twenty-one months before Apollo was ready to fly another manned mission. The command module was painstakingly reexamined and redesigned to ensure final success. All combustibles were replaced, the wiring design was completely reworked, and a quick-opening hatch was developed.

The Dream Continues

An understandable anxiety prevailed the morning of October 11, 1968. Apollo was once again poised on the launch pad with a crew aboard. Successful unmanned flights had been conducted since the accident, but failure now would almost surely shatter the dream. The long, hard months of work paid off as *Apollo 7,* the first manned Apollo mission, proved to be a resounding success. The spacecraft orbited the earth 163 times during its eleven-day voyage—traveling over 4.5 million miles—and emphatically demonstrated the integrity of the improved design. The dream of landing a man on the moon once again appeared within reach.

The next three missions, *Apollo 8* (December 1968), *Apollo 9* and *Apollo 10* (March and May 1969), followed in quick succession, each with specific mission objectives that paved the way for an actual landing on the moon. The *Apollo 8* crew actually went to the moon and completed ten lunar orbits on Christmas Eve before returning to earth. Frank Borman, James Lovell, and William Anders became the first humans to escape the earth's gravitational pull—and the first to see an *earthrise!* This seven-day round trip to the moon

The command module descends to the Pacific.

Apollo 11, "... before the decade is out."

was a severe test for all Apollo systems and its success gave the program new vigor.

Apollo 9 remained in earth orbit and conducted tests on the lunar module, the landing craft that would separate from the command module and actually land on the moon. The lunar module flew 100 miles away from the command module as extensive rendezvous and docking tests were carried out simulating the requirements of a lunar mission.

Apollo 10 repeated the course of *Apollo 8* but also carried the lunar module along. It was the final step before an actual landing would be attempted, right down to separating the lunar module and flying it to within 8 miles of the landing site before rejoining the orbiting command and service modules. It was suggested that a landing could have been attempted during the *Apollo 10* mission, but a one step at a time strategy was maintained.

The Giant Leap

On July 16, 1969, the *Apollo 11* crew of Neil Armstrong, Michael Collins, and Edwin "Buzz" Aldrin blasted off from Cape Canaveral to fulfill President Kennedy's 1961 mandate to "send a man to the moon and return him safely to earth, before the decade is out." Nearly a million people were on the beaches of Florida to witness the event, and millions more all over the earth watched on television and wished the crew Godspeed on their historic journey.

Four days later, with Michael Collins orbiting above in the command module, Buzz Aldrin and Neil Armstrong descended to the surface in the lunar module *Eagle*. On the far side of the moon the *Eagle* separated and began its slow descent to the surface, 60 miles below.

As they descended, Armstrong flew *Eagle* while Aldrin reported their progress as they neared the landing·site: "Hang tight; we're go. 2,000 feet."

The *Eagle* was slowly descending to the surface . . . 1,600 feet . . . 1,400 . . . 1,000 . . . 540 . . . 300 . . . :

Eagle: Coming down nicely . . . 200 feet [altitude] 4½ down . . . 5½ down . . . [feet per second, or about 4 miles per hour] 5 percent . . . 75 feet . . . 6 forward . . . lights on . . . down 2½ . . . 40 feet, down 2½, kicking up some dust . . . 30 feet, 2½ down, faint shadow . . . 4 forward . . . 4 forward . . . drifting to right a little . . . O.K.

Houston: 30 seconds [fuel remaining].

Eagle: Contact light! OK engine stop . . . descent engine command override off . . .

Houston: We copy you down, *Eagle.*

Eagle: Houston, Tranquility Base here, the *Eagle* has landed.

Houston: Roger Tranquility, we copy you on the ground. You got a bunch of guys about to turn blue. We're breathing again. Thanks a lot.

It was 3:18 P.M. Houston time, Sunday, July 20, 1969. Almost four hours later, Neil Armstrong became the first human to set foot on the moon, saying "That's one small step for a man, one giant leap for mankind."

A minute later he reports, "The surface is fine

Opposite: Man walks on the moon, July 20, 1969.

Right: Apollo 11 astronauts Armstrong, Collins, and Aldrin in quarantine aboard the USS *Hornet* are greeted by President Nixon shortly after their triumphant return from the moon.

and powdery. I can pick it up loosely with my toe. It does adhere in fine layers like powdered charcoal to the sole and sides of my boots. I only go in a fraction of an inch." He quickly begins photographing the surface and takes a small "contingency" soil sample.

Thirty minutes later Aldrin steps on the surface. Armstrong then unveils the engraved plaque attached to one of *Eagle's* legs. It reads:

Here men from the planet Earth first set foot upon the moon, July 1969 A.D. We came in peace for all mankind.

The Legacy of Apollo

There was celebration all around the world and ticker-tape parades throughout the United States. Yet to the person in the street, space exploration was little more than a diversion from the more pressing problems on earth. The achievement of the impossible seemed quickly forgotten. The press treated the five lunar landings following

Apollo 15 astronaut James Irwin and the lunar rover,
August 1971.

Apollo 11 as technological stunts to boost national prestige.

But surely there was more than national pride in John Kennedy's mind when he committed the nation to the goal. He must also have felt an obligation to the future of mankind—to expand horizons so that future generations might build on those achievements. In addition, he knew that a new frontier was needed to invigorate America—just as the Western frontier had invigorated America earlier in its history. JFK saw the unlimited potential of space exploration.

The Apollo program also had many practical benefits. The overriding requirement for spaceflight—that things be smaller and more lightweight—stimulated the growth of the electronics industry. Many of today's high-technology products came out of the Apollo program.

Moreover, each follow-on mission to the moon had a larger scope of objectives and explored a different area of the moon. New equipment, such as the lunar rover—a dune-buggy-like vehicle—put larger areas within reach.

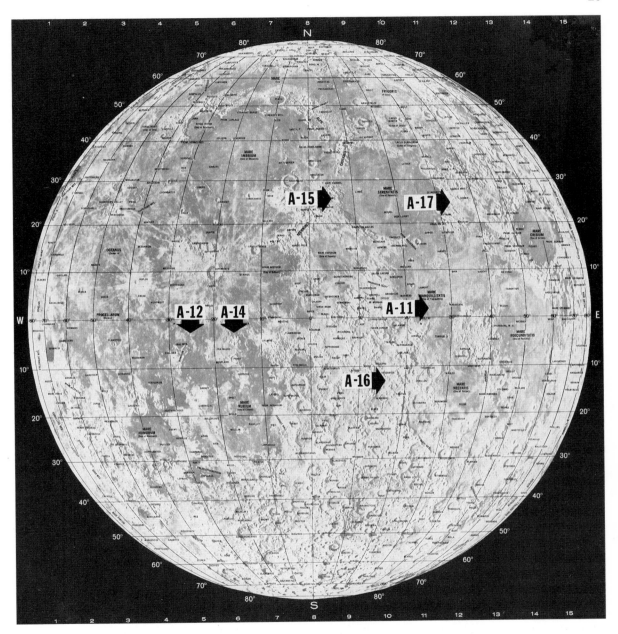

Apollo landing sites:

Apollo 11—Sea of Tranquility
Apollo 12—Ocean of Storms
Apollo 14—Fra Mauro

Apollo 15—Hadley-Apennine
Apollo 16—Descartes
Apollo 17—Taurus-Littrow

From Apollo we also learned that the moon is composed of minerals very similar to those on earth—iron, aluminum, silicon, and oxygen—and that it is especially rich in the heavy elements such as titanium.

Indeed mining the moon may one day become a necessity. And considering that it takes only one-sixth the thrust to escape the lunar gravitational field than it does to escape the earth's gravity, lunar mining could become economically attractive as the earth's resources become depleted and construction in space reaches a high level.

The Apollo program achieved six successful lunar surface explorations, boldly demonstrating the United States's scientific and technological prowess. Apollo made America the first truly space-faring nation and the exploration of the moon will forever be remembered as one of mankind's greatest achievements.

Apollo 15 astronaut James Irwin saluting the flag in this poignant photo that shows the lunar module *Falcon* and the "dune buggy." In the background, Hadley Delta rises 13,124 feet above the plain.

Apollo Missions

Mission	Date	Astronauts	Mission Duration Days:Hrs:Min	Earth Orbits	Lunar Orbits	Miles Traveled	Lunar Samples Returned
Apollo 7	Oct 11–22, 1968	Walter Schirra Donn Eisele Walter Cunningham	10:20:09	160	0	4,539,959	—
Apollo 8	Dec 21–27, 1968	Frank Borman James Lovell William Anders	06:03:00	1.5	10	578,720	—
Apollo 9	Mar 3–13, 1969	James McDivitt David Scott Russell Schweickart	10:01:00	151	0	4,208,193	—
Apollo 10	May 18–26, 1969	Thomas Stafford John Young Eugene Cernan	08:00:03	1.5	31	828,169	—
Apollo 11	Jul 16–24, 1969	Neil Armstrong Michael Collins Edwin "Buzz" Aldrin	08:03:18	1.5	30	950,598	46 lbs
Apollo 12	Nov 14–24, 1969	Charles Conrad Richard Gordon Alan Bean	10:04:36	1.5	45	950,031	75 lbs
Apollo 13	Apr 11–17, 1970	James Lovell John Swigert Fred Haise	05:22:54	1.5	Flyby	621,316	—
Apollo 14	Jan 31–Feb 9, 1971	Alan Shepard Stuart Roosa Edgar Mitchell	09:00:01	1.5	34	1,148,560	95 lbs
Apollo 15	Jul 26–Aug 7, 1971	David Scott Alfred Worden James Irwin	12:07:11	1.5	74	1,272,187	170 lbs
Apollo 16	Apr 16–27, 1972	John Young Thomas Mattingly Charles Duke	11:01:51	1.5	64	1,388,429	209 lbs
Apollo 17	Dec 7–19, 1972	Eugene Cernan Ronald Evans Harrison Schmitt	12:13:51	2	75	1,482,721	258 lbs

CHAPTER 4

THE DECADE OF THE 1970s

"THE SKYLAB MISSION WILL PROVIDE MAN HIS FIRST OPPORTUNITY TO STUDY —FROM THE VANTAGE VIEW OF EARTH ORBIT—THE SUN, THE EARTH, HIMSELF, AND TECHNOLOGY."

SKYLAB PRESSBOOK, 1973

In February 1969, at the height of the Apollo program, President Nixon appointed a Space Task Group to make definitive recommendations concerning the direction the U.S. space program should take in the post-Apollo period. The Space Task Group recommended that increased emphasis be placed on science, and on developing technologies in the fields of communications and meteorology.

The task group recommended the development of a new type of space transportation system, one that would be *reusable* and provide a major improvement in operational capability. The recommendation to develop reusable space trans-

Skylab, the first U.S. space station, as viewed by the approaching Apollo command module. Notice the parasol covering the exposed workshop area.

portation arose from the inefficiency of existing rockets. Once used, a rocket either fell into the ocean and sank, or floated aimlessly in space. Detailed studies conducted between 1970 and 1971 to identify a more efficient method of spaceflight led to the design of the space shuttle.

In January 1972 President Nixon initiated the development of the space shuttle saying:

> The United States should proceed at once with the development of an entirely new type of space transportation system designed to help transform the space frontier of the 1970s into familiar territory, easily accessible for human endeavor in the 1980s and 1990s. . . . In short, this system will go a long way toward delivering the rich benefits of practical space utilization and the valuable spin-offs from space efforts into the daily lives of Americans and all people.

Skylab—Investigating Earth and Space

It would be nine challenging and often frustrating years before the first space shuttle mission.

While the shuttle was being designed and constructed, the space program focused on other means of exploration. Skylab, the first American space station, was the major U.S. manned space initiative in the early 1970s. Skylab's ambitious mandate was to serve as a long-duration orbiting space station, housing scientific, technological, and biomedical studies.

Despite all the well-laid plans, Skylab came very close to failure on its first day in space. During

Astronaut Owen Garriott stationed at the Apollo telescope mount console aboard the Skylab space station.

the launch of the unmanned workshop on May 14, 1973, a micrometeoroid shield tore loose and damaged the main solar array that was to supply Skylab's electricity. Skylab's power-generating capacity was cut by over 50 percent. Even more potentially damaging—the loss of the micrometeoroid shield exposed the workshop to soaring temperatures that threatened to destroy the $200 million spacecraft.

The following day—May 15, 1973—an Apollo spacecraft was to have been launched with a three-man crew aboard to dock with the orbiting Skylab. The Apollo launch was delayed as mission planners assessed the combined electrical and heating problems aboard Skylab and evaluated the options.

Eleven days later—May 25, 1973—Pete Conrad, Joe Kerwin, and Paul Weitz became the first repair crew in space. They deployed an umbrella-like sun shield over the exposed areas of Skylab, thus lowering the temperature inside the workshop to a comfortable 75°F. Several days later they spent four hours freeing the jammed solar array—restoring about 3000 watts of desperately needed electricity and enabling the crew to complete their full twenty-eight-day mission.

Two follow-on Skylab missions in mid-1973 and early 1974 built on these accomplishments. The missions demonstrated that working in space for long periods of time is practical, and that unique experiments could be performed. For example, it had long been theorized that processing materials in zero gravity would have advantages over earth-based processing. Skylab experiments yielded an understanding of the fundamental behavior of materials processed in zero gravity and identified many potential space products, difficult or impossible to produce on earth. These basic fundamentals, combined with subsequent ground-based research, are being applied to shuttle experiments today.

The three Skylab missions of twenty-eight, fifty-nine, and eighty-four days allowed extensive study of the earth as well as unique investigations of our solar system with telescopes and cameras unobstructed by the earth's atmosphere. More than 180,000 photos of the solar system were taken during the three missions. Earth survey experiments included over 40,000 photos used to improve weather prediction, crop forecasting, and water and forestry planning.

Overall, the Skylab program played a major part in the transition from the Apollo era of the 1960s to the shuttle era of the 1980s and beyond.

Skylab Missions

Mission	Date	Astronauts	Mission Duration Days:Min:Sec	Earth Orbits	Miles Traveled	Launch Vehicle
Skylab 1	May 14, 1973	Unmanned	—	—	—	Saturn V
Skylab 2	May 25–Jun 22, 1973	Charles Conrad Paul Weitz Joe Kerwin	28:00:49	404	11,470,000	Saturn 1B
Skylab 3	Jul 28–Sep 25, 1973	Alan Bean Jack Lousma Owen Garriott	59:11:09	858	24,490,000	Saturn 1B
Skylab 4	Nov 16, 1973–Feb 8, 1974	Gerald Carr Bill Pogue Ed Gibson	84:01:16	1214	35,030,000	Saturn 1B

The Meeting in Space

Apollo and *Soyuz*—East meets West 140 miles above the earth.

There would be one other U.S. manned mission in the 1970s, the Apollo-Soyuz Test Project (ASTP). In the spirit of detente, President Nixon and Soviet Premier Kosygin took the bold step of proposing the first international manned space mission. In July 1975, U.S. astronauts Tom Stafford, Vance Brand, and Deke Slayton, and Soviet cosmonauts Aleksey Leonov and Veleriy Kubasov, culminated three years of combined effort when they met in space for two days of research and international goodwill.

Above and beyond the political aspects of the ASTP, its major objective was to see if international spacecraft could rendezvous and dock for emergency space rescues. The docking mechanism was itself symbolic of the joint mission. Though jointly designed, each craft had been developed and built in its respective country; though different in detail, they both worked together.

During the two days the spacecraft were docked, the crews conducted numerous joint experiments—overcoming language, operating, and hardware differences—thereby proving the feasibility of international cooperation in space. Furthermore, scientific experiments conducted during ASTP enhanced the prospects for future international space cooperation. The combined efforts of NASA and the Soviet Academy of Sciences led to an exchange of scientific knowledge between the two countries that continued for the remainder of the decade.

The ASTP mission represented the fifteenth and final Apollo command module to carry men into space. It was also the last manned spacecraft flown by U.S. astronauts. It would be almost six years before another American would venture into space. The next mission would use the *reusable* space shuttle.

The Space Transportation System

The space shuttle, or as it is more formally referred to, the Space Transportation System (STS), has been described as a space-going "truck" for hauling cargoes into orbit and return-ing them safely to earth. In a sense it is a "truck," but it is also the most sophisticated and versatile spacecraft ever developed.

Operating as much more than a transportation system, the space shuttle has the combined abil-ities of an aircraft, a launch vehicle, and a spacecraft. The shuttle is comprised of four major components: 1) the airplane-like orbiter, designed to fly over one hundred missions; 2) three main engines, the first reusable rocket engines ever built; 3) the large external tank, which supplies

Tom Stafford and Aleksey Leonov in tunnel connecting the *Apollo* and *Soyuz*.

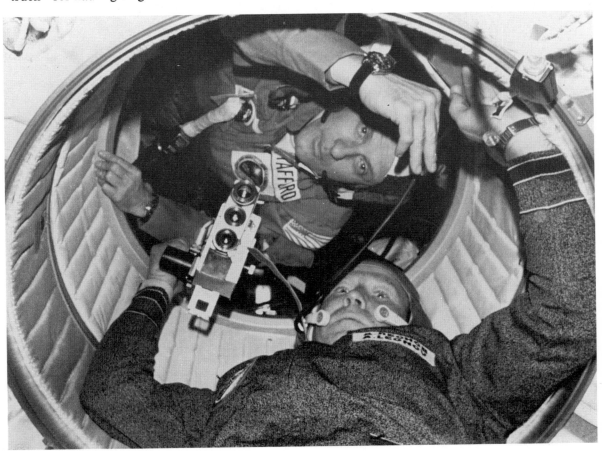

fuel to the main engines; and 4) two solid rocket boosters, which combined with the main engines, produce almost 7 million pounds of thrust to boost the shuttle into orbit. Of these four components only the external tank is expendable.

The space shuttle was the logical next step in space transportation. We had evolved from the very small Mercury spacecraft to the immense Apollo and Skylab vehicles. Now it was time for a more versatile spacecraft that would make space-flight routine.

The space shuttle can be accurately described in the following manner: it is launched as a rocket, orbits the earth as a spacecraft, and returns as a glider to a runway landing. Since the orbiter returns powerless from space, it has only one attempt at landing. To thoroughly understand the shuttle system's performance capabilities two distinct series of flight tests were undertaken. First came the Approach and Landing Test program (ALT). Following ALT, four orbital flight tests would certify the shuttle to carry commercial and scientific cargo into space.

Evolution of manned spacecraft: Mercury, Gemini, Apollo, Skylab, Apollo-Soyuz, and the shuttle.

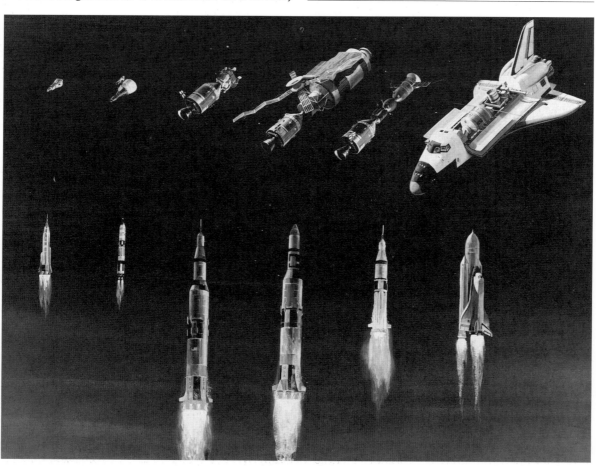

Vital Shuttle Statistics

Dimensions

The orbiter is 122 feet long with a wingspan of 78 feet. It is about the same length as a Boeing 737, but with a wider midbody. The solid rocket boosters are 149 feet tall (16 stories). The external tank reaches 184 feet above the launch platform (21 stories).

Weight

The orbiter alone weighs approximately 210,000 pounds. The orbiter, solid rocket boosters, and external tank weigh 4.5 million pounds at launch.

Cargo Capability

In its 15-foot-by-60-foot payload bay the shuttle can carry up to 65,000 pounds of cargo into a due-east orbit out of Kennedy Space Center. When launched into a polar orbit out of Vandenberg Air Force Base the shuttle can put 32,000 pounds into orbit. In addition to its superior launch capability, the shuttle can also *return* 32,000 pounds of cargo to earth.

Propulsion

Each solid rocket booster produces 2.9 million pounds of thrust and the three main engines generate 375,000 pounds each. The total, at launch, is over 6.9 million pounds of thrust.

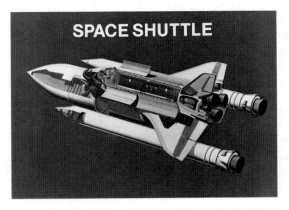

SPACE SHUTTLE

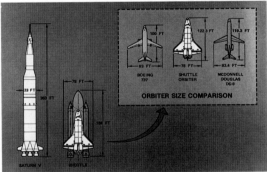

Above: The overall shuttle system: orbiter, external tank, and twin solid rocket boosters.

Below: The shuttle compared to a Saturn V, 737, and DC-9.

Fuel Consumption

The orbiter's engines use a total of 64,000 gallons of fuel and oxydizer per minute. The external tank holds 380,000 gallons of liquid hydrogen and 140,000 gallons of liquid oxygen. The boosters' solid fuel is mostly ammonium perchlorate, aluminum powder, and iron powder. Each booster has 1.1 million pounds of fuel.

Shuttle orbiter *Enterprise* released over Edwards Air Force Base during Approach and Landing Test, August 1977.

The ALT flights were conducted at Edwards Air Force Base in California, using a modified Boeing 747 as a carrier vehicle. The orbiter rode piggyback atop the 747 to an altitude of 25,000 feet where it was released as a glider to evaluate its flight and landing characteristics. In 1977 five successful ALT flights were conducted during the months of August, September, and October using the *Enterprise,* a prototype orbiter. Meanwhile, *Columbia,* the first space-worthy orbiter, was well under construction. *Columbia* was delivered to the Kennedy Space Center in early 1979. Although a substantial amount of work remained, its initial launch date was expected in approximately one year.

Building a Reusable Spacecraft

A revolutionary concept in spaceflight, the shuttle has been one of the boldest engineering endeavors ever undertaken. New technologies were demanded to make the reusable shuttle easy to maintain. Airline industry techniques were applied to ensure that it would in fact become the world's first "spaceline."

Two of the most difficult challenges encountered were the development of the space shuttle main engines (SSME) and the orbiter's thermal protection system (TPS). Construction of the main engines—designed for reuse up to fifty-five times—required an ease of maintenance similar to that of commercial jetliners.

The SSMEs are compact enough to fit inside the tail end of the orbiter, yet they must deliver 375,000 pounds of thrust each. The combustion chamber pressure had to be increased to 3,000 pounds per square inch (psi), more than quadruple the 700 psi of the Apollo engines. Furthermore, the turbo pumps that deliver the fuel to the combustion chamber weigh little more than a truck

42

engine, yet they produce as much as 77,000 horsepower.

The orbiter's thermal protection system is like an outer skin that protects the orbiter's aluminum structure from the scorching 2,700°F heat of reentry and the severe −250°F environment of space. Earlier thermal protection systems on the Apollo, Gemini, and Mercury spacecraft were made of a heavy ablative material that slowly burned off

Shuttle main engines being prepared for orbiter installation.

during reentry. The orbiter's enormous size and requirement for reuse called for the development of a new lightweight material that could be easily maintained. The shuttle's TPS is composed of over 31,000 tiles of almost pure silica that dissipate the furnacelike heat of reentry while remaining intact. *Each* of these tiles required a unique shape matching the contour of the orbiter's aluminum outer skin.

As is often the case with large-scale projects breaking new technological ground, things didn't always go smoothly. Engines blew up during testing. Thermal protection system tiles fell off. These kinds of problems were inevitable—considering the size of the shuttle program and the complexities involved.

Moreover, many of the shuttle's early problems can be traced to inadequate funding. The shuttle was developed on a relatively tight budget. For example, earlier programs were developed with unmanned test flights prior to actual flight. There would be no unmanned shuttle test flights.

Because of funding limitations, NASA was forced to develop and test the shuttle on a "success-oriented" program. Otherwise, the shuttle might never have been approved at all. As it is, the shuttle is a compromise between proposed design and fiscal limitations. Originally, it was conceived as a *fully reusable* two-stage vehicle with even greater cargo capability than the present vehicle. Success-oriented testing meant that engines would be tested as assembled units instead of individual components (as with earlier programs). This type of testing greatly increased the level of risk—especially considering all the technological ground the SSMEs were breaking.

As for the TPS, the tiles were originally thought to be little more than a coating for the orbiter's aluminum skin. It wasn't until *Columbia* had over 20,000 tiles on her, that structural testing and loads analysis showed the complexity of the system. The size of the tiles vary from less than 1/2-inch to 3½ inches in thickness, and from about 1-inch to about 6-inches square. The structural testing and loads analysis showed that each individual tile was a subsystem in itself—calling for much more than simply attaching material to a structure. This type of testing should have been conducted years earlier. Since it wasn't, each indi-

Orbiter right after reentry. Inserts show diversity of thermal protection tiles.

vidual tile had to be pull-tested to meet the new loads requirements. This tedious task slowed *Columbia's* progress by a full year.

Despite the fiscal limitations, technical challenges, and hybrid design (partly reusable instead of fully reusable), all obstacles were overcome and *Columbia* stood poised on the launch pad in April 1981.

The shuttle as it might have been and the current vehicle.

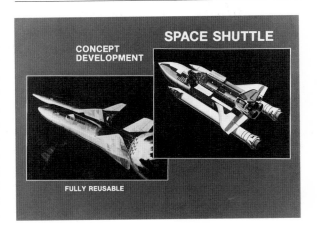

CONCEPT DEVELOPMENT

SPACE SHUTTLE

FULLY REUSABLE

CHAPTER 5

THE SHUTTLE OPENS A NEW ERA

"WITH THE SPACE SHUTTLE, WE TESTED OUR INGENUITY ONCE AGAIN, MOVING BEYOND THE ACCOMPLISHMENTS OF THE PAST INTO THE PROMISE AND UNCERTAINTY OF THE FUTURE."

PRESIDENT RONALD REAGAN
APRIL 1981

On the morning of April 12, 1981, the first space shuttle mission (STS-1) ushered in a new era of space transportation. Originally scheduled for April 10, a two-day delay due to technical problems led to an incredible coincidence: it had been *exactly* twenty years since Soviet cosmonaut Yuri Gagarin became the first man to orbit the earth on April 12, 1961.

As *Columbia* rose from its launch pad, it carried with it the hopes and dreams of thousands of people across the nation and millions of others all over the world. Never before had a vehicle ventured into orbit to return as an airplane. No longer would we see splashdowns in the ocean—from

A new era begins, April 12, 1981.

now on it would be touchdowns on a runway!

From its awe-inspiring launch to its perfect touchdown, *Columbia*'s performance was praised. The astronauts flying *Columbia*, John Young and Bob Crippen, hailed her as an incredible flying machine and the first true spaceship after they touched down at Edwards Air Force Base, with a landing smoother than many a commercial jet. Crippen later commented, "From a pilot's standpoint you could not ask for a more superb flying machine." That first shuttle mission was a shot in the arm for the U.S. space program and for the nation as a whole; it represented a technological achievement of the highest order and demonstrated that American know-how was indeed alive and well.

The timing couldn't have been better. The nation was in the midst of a recession and had just suffered through two years of worldwide humiliation. Iran had held fifty-two hostages at the American Embassy in Tehran for over four-hundred days and embassies in other nations had been attacked by terrorists. *Columbia*'s successful first flight, STS-1, simply made Americans feel good about their country again, about America's ability to pursue and achieve large endeavors.

45

In three subsequent missions—November 1981, March and June 1982—the shuttle completed its orbital flight testing and lived up to its advance billing of being the world's first reusable spacecraft. The shuttle not only exceeded performance expectations during these first four missions, but it also proved to be an excellent platform for conducting research in space.

The second mission, STS-2, carried the first shuttle cargo, designated OSTA-1 (NASA's Office of Space and Terrestrial Applications' first payload). OSTA-1 consisted of seven scientific experiments on subjects including atmospheric, ocean and earth resources. STS-2 also successfully tested the shuttle's Canadian-built robot arm which has since been used to deploy and retrieve satellites.

STS-3 carried the first payload for NASA's Office of Space Sciences, designated OSS-1.

OSS-1 consisted of nine experiments encompassing a wide variety of scientific disciplines including astronomy, space plasma physics, and solar physics. This mission also demonstrated the shuttle's flexibility when heavy rains turned the designated landing site at Edwards Air Force Base into a small lake. *Columbia* remained in orbit an extra day and made a perfect landing the following morning at the White Sands Test Facility in New Mexico (since renamed White Sands Space Harbor).

Columbia's first four flights successfully demonstrated the practicality of reusable spacecraft and forever revolutionized spaceflight. Each flight crew praised their craft's performance as they proved its ability to make spaceflight routine. President Reagan accurately described the shuttle when he attended its fourth landing on July 4, 1982. The president compared the fourth mission,

No more spashdowns in the ocean. April 14, 1981.

which completed the shuttle's test flights, to the golden spike that marked the completion of the first transcontinental railroad. The railroad opened up the American West—the shuttle is performing the same role in space. During his address the president also referred to the shuttle's capability to establish a permanent presence in space, an allusion to NASA's plans for a space station.

The Operational Phase Begins

Having passed its orbital flight tests with flying colors, the shuttle was ready to carry cargo for paying customers. Beginning with its fifth mission (November 11-16, 1982), the shuttle entered its operational phase. The first four-man crew carried out the seven-day mission: commander Vance Brand, pilot Bob Overmyer, and mission specialists Joe Allen and Bill Lenoir. Their assignment was to deploy the shuttle's first commercial satellites.

The two large satellites are owned by Telesat of Canada (the Canadian phone company) and Satellite Business Systems (SBS) of McLean, Virginia. The satellites were approximately 9 feet high and 5 feet in diameter when stowed in the orbiter's cargo bay. After reaching their operational orbits, their solar panels were fully deployed to a height of 21 feet.

Both satellites were carried in cradles equipped with springs that ejected them out of the cargo bay into a drift orbit. Forty-five minutes later, once the orbiter was safely away from the satellite, the satellites' own engines were fired to take them from the shuttle staging point 180 miles above the earth to the satellites' operational orbit at 22,300 miles. From this 180-mile orbit, away from the earth's gravitational pull, satellites require only a small additional boost to reach their final orbit.

Above: The 50-foot shuttle arm passes its first test with flying colors, STS-2, November 1981.

Below: The multidiscipline OSS-1 experiment package, STS-3, March 1982.

(The shuttle will launch hundreds of satellites from this staging point). Today both satellites are in operation providing communications services to the United States and Canada. STS-5 commander Vance Brand accurately summed up their mission with two simple words, "We deliver!"

"The fourth landing of the *Columbia* is the historical equivalent to the driving of the golden spike which completed the first transcontinental railroad. It marks our entrance into a new era . . . [the ships] are ready to provide economical and routine access to space for scientific exploration, commercial ventures, and for tasks relating to national security. Simultaneously, we must look aggressively to the future by demonstrating the potential of the shuttle and establishing a more permanent presence in space."

STS-6: *Challenger* Comes On Line

Challenger, the second shuttle orbiter, came into service with the sixth shuttle flight in April 1983. STS-6, a six-day mission, carried the first of six huge tracking and data relay satellites (TDRS) that will be launched into orbit throughout the 1980s. The enormous TDRS satellites are the largest and most capable communication satellites ever built. They measure an imposing 52 feet by 46 feet when fully extended. The TDRS system is a "switchboard in the sky" designed to simultaneously connect up to thirty-two satellites and the shuttle with their ground stations. They will handle the growing amounts of data that are being routed through space. Each TDRS satellite can transmit a stream of 300 million bits of data per second (equal to ten sets of a fourteen-volume World Book encyclopedia). These satellites will revolutionize space communications and replace many of NASA's satellite-tracking facilities around the world. The majority of NASA's ground-based facilities are twenty years old and can only track satellites as they pass overhead. The TDRS, on the other hand, has the ability to track satellites from above on a nearly continuous basis.

Above: The first two shuttle-deployed satellites being prepared for flight: the SBS satellite in the foreground and the extended Telesat behind it.

Opposite: SBS begins second leg of the journey, STS-5, November 1982.

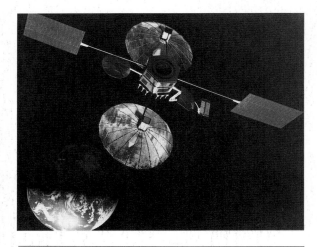

Above: TDRS, the switchboard in the sky.

Below: Astronaut Story Musgrave testing some tools 175 miles above the earth. STS-6, April 1983.

The first shuttle extravehicular activity (EVA) or spacewalk was also conducted during the STS-6 mission. The three-hour EVA tested the astronauts' mobility in their new space suits, and their ability to operate the numerous tools for the repair and replenishment of satellites. It was the first

U.S. EVA in nine years and it paved the way for routine spacewalks on follow-on shuttle flights. With *Challenger* joining the operational *Columbia,* the shuttle fleet began to take shape.

STS-7: Oh What a Ride!

Challenger's second voyage will long be remembered for Sally Ride, America's first woman astronaut, but this voyage also demonstrated how space is rapidly becoming international. During the seven-day mission the five-person crew deployed two communication satellites (one each for Canada and Indonesia), the experimental shuttle pallet satellite (SPAS) for Germany, and they retrieved SPAS with the Canadian robot arm.

The SPAS experiments were two-fold in nature, conducted first in the shuttle, then with the satellite in orbit. During the first three days the crew activated SPAS-mounted experiments for Germany as SPAS remained in the payload bay where it conducted materials processing experiments and

Sally Ride during her historic mission.

Unique remote photo taken from the orbiting shuttle pallet satellite (SPAS)—the first time the orbiter was viewed from this spectacular perspective.

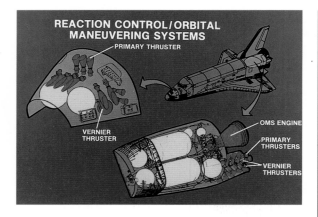

The orbiter's propulsion units give it flexibility to maneuver in orbit and point in any direction.

Shuttle Orbital Maneuverability

Two sets of smaller engines are used once the shuttle is in orbit. The orbital maneuvering system (OMS), the two engines located just above the main engines, is used for orbital maneuvers. A second set of engines, the reaction control system (RCS), is used for smaller attitude changes. The RCS is a series of forty-four small thrusters (thirty-eight primary, six vernier) that can point the orbiter in any direction necessary. Where the OMS is primarily used for major orbit changes, the RCS is used to change the position of the orbiter. For example, depending on the requirements of its cargo, the orbiter can be pointed toward the earth for earth-observation experiments, or out into deep space for astronomy experiments.

photographed the earth. The real fun began on the fourth day when SPAS was deployed and played a game of "orbital tag" with the shuttle. The SPAS was deployed and retrieved twice as the crew demonstrated the shuttle's ability to track, rendezvous, and capture a satellite in orbit (practice for upcoming satellite-repair missions). The STS-7 crew made light of their accomplishments as they proclaimed, "We pick up and deliver!"

Guy Bluford photographing an experiment during STS-8.

STS-8: *Challenger* Roars Into the Night

A spectacular instant sunrise could be seen for 400 miles as *Challenger* thundered off the launch pad at 2:32 A.M. on August 30, 1983. The eighth shuttle mission was a combination of firsts. It was the first night launch *and* landing, and its five-person crew included the first black U.S. astronaut, Guy Bluford.

The night launch was necessary to deploy an Indian satellite while over its ground station in India for tracking purposes. The night landing was important because future night landings might be necessary in emergencies or may even be desirable when landing in Florida where the weather is often better at night than during the day.

India's satellite *Insat 1-B* was successfully deployed and is providing improved telephone, television, and weather service to remote parts of India. Satellite communications are vital to India, whose population of over 700 million people is dispersed over a huge geographical area.

Later in the mission the crew put the robot arm through its toughest tests yet as they used it to berth and unberth a four-ton practice satellite the size of a pick-up truck. The robot arm tests were for follow-on shuttle missions when the arm would be used to capture orbiting satellites for repair.

Finally, STS-8 successfully tested several shuttle antennas by communicating with mission control through the tracking and data relay satellite which was launched during STS-6. This satellite will be used on all future shuttle missions to greatly improve the shuttle's ability to relay information to the ground.

The "instant sunrise" of the first shuttle night launch, August 30, 1983.

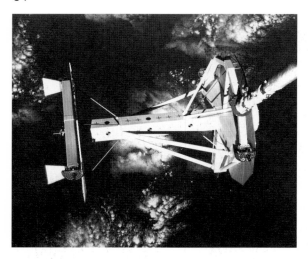

STS-9: The Debut of Spacelab

The first Spacelab mission ushered in a new chapter in international space cooperation. Built by the European Space Agency, a consortium of eleven European nations, the Spacelab is a versatile, general-purpose laboratory which is carried in the orbiter's cargo bay.

Skylab's first mission witnessed the most intense scientific experimentation ever conducted in space. During the ten-day mission Spacelab gathered fifty *times* as much data as Skylab did over its

Above: Shuttle arm captures a 4-ton ''practice satellite,'' STS-8, September 1983.

Below: West German astronaut Ulf Merbold conducting an experiment aboard *Spacelab-1.*

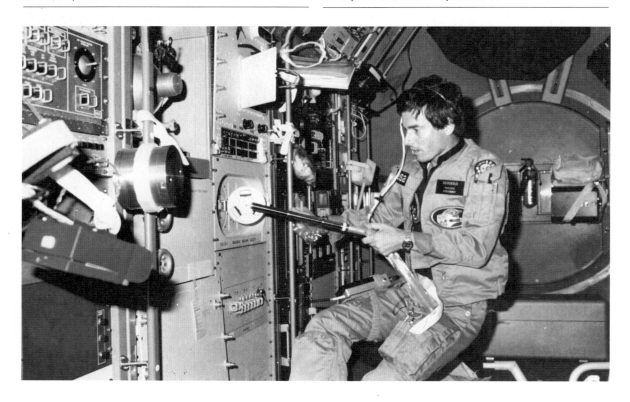

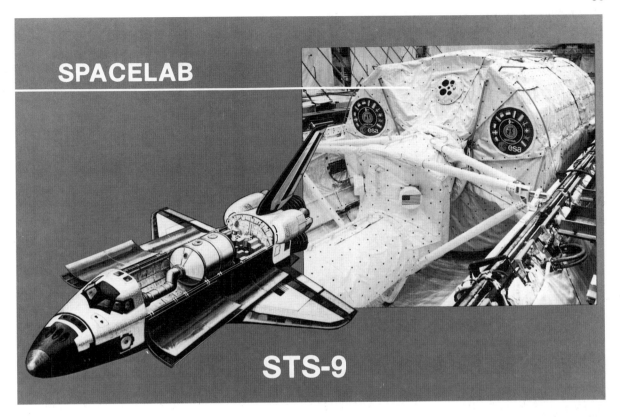

SPACELAB

STS-9

172 days of manned operation from 1973 to 1974. Scientists from Europe, Japan, Canada, and the United States provided seventy-seven experiments covering five research areas: astronomy, space plasma physics, atmospheric physics, life sciences, and materials processing in zero gravity. Spacelab's bounty of data will keep scientists busy for years.

In addition, Ulf Merbold of West Germany became the first European astronaut to fly aboard the shuttle. Merbold was the first of many foreign astronauts who will fly aboard the shuttle. International cooperation in space will grow in the future as astronauts from Great Britain, France, the Netherlands, Canada, Mexico, India, Saudi Arabia, Indonesia, and Japan fly on shuttle missions.

Artist's concept of *Spacelab-1* installed in the cargo bay; inset shows the actual hardware.

STS-11: A Human Satellite

There would be five shuttle missions in 1984 —each one breaking new ground. STS-11 brought Buck Rogers right out of the science-fiction pages and put him in the news as astronauts Bruce McCandless and Bob Stewart put the shuttle's flying backpack through tests for an upcoming satellite retrieval and repair mission.

The Manned Maneuvering Unit

The manned maneuvering unit (MMU), a backpack propulsion device, is used when astronauts have to venture beyond the payload bay for such activities as satellite servicing and repair, in-space construction, and rescue operations. Its twenty-four thrusters give the astronauts the precise control needed when approaching sensitive satellite components. They can hold a station-keeping position, or they can maneuver up, down, sideways, backwards, and forward.

The unit can operate for up to six hours at a time and can be refueled for longer operations by plugging into the orbiter's nitrogen fuel supply. Two MMUs are currently in operation.

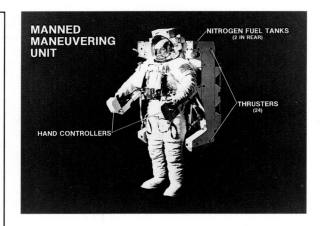

Buck Rogers comes out of the comic books.

Bruce McCandless, "The Human Satellite," flies the manned maneuvering unit for the first time.

Each astronaut flew the backpack, known as the manned maneuvering unit, 150 feet and then 300 feet from the orbiter. It was a startling sight to see a man alone in the cold void of space without a safety tether attached—in effect making him a human satellite.

After running the backpack through its flight tests, the astronauts returned to the payload bay to test satellite repair tools and the manipulator foot restraint. The manipulator foot restraint is similar to a cherry picker. It serves as a tool carrier and work platform as the astronaut stands in it on the end of the robot arm leaving his hands free as the robot arm moves him around the payload bay.

Unfortunately, STS-11 wasn't entirely successful. Two satellites deployed on the mission, *Westar VI* and *Palapa B-2,* left the payload bay smoothly but their upper stages malfunctioned and left them in useless, 560-mile orbits. NASA would not find a way to retrieve the two errant satellites until their spectacular rescue nine months later (see STS-19).

Opposite: McCandless on the manipulator foot restraint at the end of the shuttle's robot arm.

STS-13: The Ace Satellite Repair Company

With the successful demonstration of the manned maneuvering unit, NASA was now ready for the real thing: retrieval and repair of a satellite in orbit, and Solar Max was the candidate. The Solar Maximum satellite had been launched in February 1980 to study the sun during the height of its eleven-year cycle. Solar Max had performed flawlessly until December 1980 when both its main electronics box and the fuses in its attitude control module failed. Without attitude control, Solar Max was unable to conduct the fine-pointing solar observation it was designed for.

The STS-13 plan was to have astronaut George Nelson use the MMU to rendezvous with, dock with, and stabilize the slowly spinning Solar Max. The shuttle's robot arm would then place the satel-

---

George Nelson attempting to dock with Solar Max.

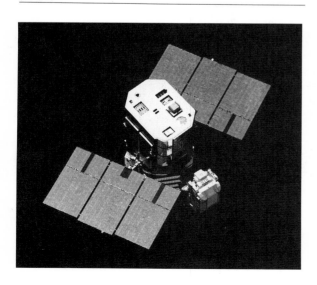

lite in the cargo bay for repair. Everything was going smoothly until Nelson reached Solar Max and found an unexpected 1/2-inch grommet on the satellite that wouldn't allow his docking tool to operate. After three unsuccessful docking attempts, Nelson tried to stabilize the 5,000-pound satellite with his hands. This only made things worse as Solar Max began to wobble. Nelson returned to *Challenger* with the payload bay still empty.

It was crisis time. Workers at mission control in Houston and NASA's Goddard Space Flight Center in Maryland worked around the clock looking for alternatives. Finally, after thirty-six hours of frantic effort, Goddard (the NASA center responsible for Solar Max since its launch) came up with a plan that worked. By reprogramming Solar Max from the ground they were able to regain control of the tumbling satellite so that the orbiter's robot arm could grapple it and place it in the cargo bay.

It was a major success for the Goddard team: in another hour Solar Max probably would have been lost. The tumbling had taken Solar Max's power-producing solar panels out of the sun's rays 80 percent of the time, and its batteries were almost dead.

With Solar Max safely in the payload bay, Nelson and James van Hoften were able to replace the faulty attitude control module, main electronics box, and place a cover over the satellite's X-ray polychromator. Only thirteen days after being repaired, Solar Max observed the largest solar flare seen on the sun since 1978. The age of the throwaway satellite was over.

STS-14: *Discovery* Joins the Fleet

Discovery made its maiden voyage in August 1984, becoming the third orbiter to venture into space. The mission established several firsts.

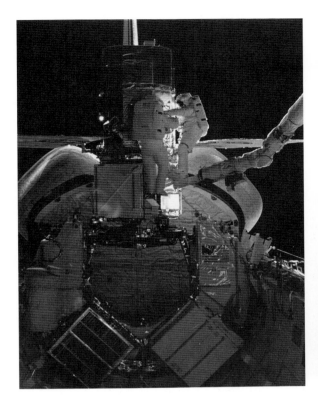

STS-14 was the first to carry three satellites on the same mission. Two of the satellites used upper stages similar to those that failed on STS-11 (they worked this time), and the third satellite was the first of the enormous Leasat satellites. The 14-foot-wide Leasats are designed specifically to take advantage of the shuttle's 15-foot-wide cargo bay. Five Leasats will be launched by the shuttle to provide communications for the U.S. Navy. STS-14 carried the first commercial astronaut —Charles Walker, an employee of McDonnell Douglas who conducted a promising pharmaceutical processing experiment (discussed further in chapter 6) and Judy Resnik, America's second woman astronaut, who made her first flight on this mission.

STS-14 also carried a 105-foot-long deployable-retractable solar array that has applications for

Left: James van Hoften and George Nelson repairing Solar Max.

Above: The 14-foot-wide "shuttle optimized" Leasat.

future satellites and space stations. Because the array is so thin (8/1000 of an inch) the entire 105-foot array could be folded up like an accordion into a box only 7 inches deep. Earlier solar arrays (such as Skylab's) were rigid, twice as expensive, and weighed eight times as much. The large solar array was extended and retracted several times to test its mechanical and dynamic properties. Such an array will not only find applications on satellites and space stations, but could also be used to supply electricity to the shuttle to extend its orbital stay time beyond the typical seven-day mission.

STS-17: An Earth-Oriented Space Mission

On *Challenger's* October 1984 mission the shuttle broke even more ground by carrying a crew of seven: Sally Ride, making her second flight; Kathy Sullivan, on her first; Marc Garneau, the first Canadian in space; Paul Scully-Power, an oceanographer for the U.S. Navy; Bob Crippen, making his fourth shuttle flight, pilot Jon McBride; and mission specialist Dave Leestma.

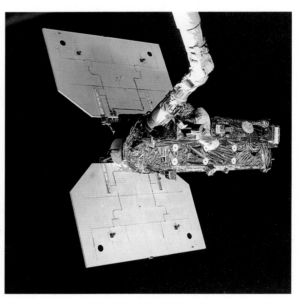

Left: Solar array demonstration—the 13-foot-wide array is extended to its full 105-foot length.

Above: Earth radiation budget satellite (ERBS) as it was deployed by the shuttle's robot arm.

Opposite: Dave Leetsma *(left)* and Kathy Sullivan during satellite refueling demonstration.

Three payloads on this mission were strictly earth-oriented: the large format camera, which took high resolution photographs of the earth's surface; the shuttle imaging radar-B, an enhanced version of the synthetic aperture radar that flew on STS-2 (discussed in detail in the next chapter); and the earth radiation budget satellite (ERBS).

ERBS is a major element of the nation's national climate program. Working together with two weather satellites, ERBS will gather data for long-term weather prediction by measuring the earth's radiation budget—the thermal equilibrium that exists between the sun, the earth, and space. The radiation budget is in constant flux as energy is received from the sun, transported by the oceans and atmosphere, and radiated back to space.

To cap off the mission, Kathy Sullivan and Dave Leestma conducted a satellite refueling demonstration in the cargo bay, making Sullivan the first American woman to walk in space, and demonstrating the shuttle's unique satellite-refueling capability. That capability will be put to use in the future as satellites are refueled to extend their useful lives.

STS-19: Two Up, Two Down

Discovery returned to space in November 1984 for its second mission—an especially challenging one. Two satellites would be deployed, one for Canada and another Leasat for the U.S. Navy. In addition, two satellites that had been placed in useless orbits during STS-11 *(Westar VI* and *Palapa B-2)* were to be retrieved.

Unlike Solar Max (which was designed to be retrieved), these two satellites didn't have any grappling points—the points had to be invented. The result was a "stinger" protruding from astronaut Dale Gardner's midsection, a device designed to clamp onto the attach fitting where the satellite's upper stage had been located. The

stinger also had a grapple fixture so the shuttle's robot arm could retrieve it and return the captured satellite to the cargo bay where it could be secured.

Unlike Solar Max, these two satellites could not be repaired in space and had to be returned to earth for refurbishment. Stowing the satellites in the cargo bay wasn't as easy as expected. A bracket that would have allowed the robot arm to hold the satellite while the astronauts secured it for stowage wouldn't fit. Joe Allen had to hold the 1,200-pound satellite for an entire orbit while Gardner secured it.

The two satellites were safely returned to earth and became the first used satellites for sale. The successful return prompted Lloyd's of London to award silver medals to Allen and Gardner for their "extraordinary exertions to preserve property from perils."

Below: Joe Allen holds 1,200-pound satellite as Dale Gardner stows the "stinger."

Opposite: Dale Gardner, with "stinger," approaches *Westar VI* over the Bahamas.

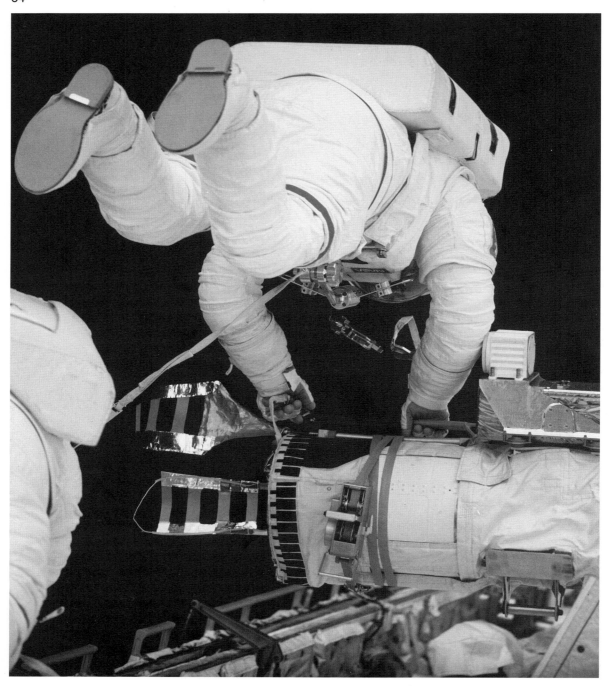

STS-20: The Pentagon's First Dedicated Mission

STS-20 was the first mission to carry only military payloads. In spite of the fact that it was a secret mission, the press widely reported that the shuttle's "secret" cargo was a signal intelligence spy satellite for monitoring Soviet electronic messages. The Pentagon tried to discourage speculation about the satellite itself without much success but did report that the upper stage successfully placed the satellite into a 22,300-mile geosynchronous orbit. In geosynchronous orbit the satellite's orbit matches the earth's orbit so that the satellite remains over the same ground location at all times. Many types of satellites operate from that orbit. It could indeed have been a signal intelligence satellite, but it could also have been a communications or weather satellite.

The Pentagon will account for one-third of future shuttle missions—up to eight per year beginning in 1988.

STS-23: A Valiant Rescue Attempt

STS-23 seemed jinxed even before it was launched. The mission was delayed no less than five times due to a series of mishaps. First there were problems with one of the satellites on board, followed by a decision to "roll back" from the launch pad to destack the shuttle and change the cargo. Then while *Discovery* was in the orbiter processing facility, a 2,500 pound overhead work platform fell on one of the payload bay doors resulting in a six-week delay.

Jeff Hoffman attaching "flyswatters."

Finally, on April 12, 1985, the mission began with another flawless launch. The crew quickly got to work and on their seventh orbit they successfully deployed another Telesat satellite for Canada, the fourth Canadian satellite deployed by the shuttle.

The second day started out fine as the crew deployed another of the huge Syncom satellites for the U.S. Navy. The Syncom left the cargo bay as smoothly as the two others deployed on earlier missions. Then nothing happened. The satellite was supposed to deploy a small antenna for ground communications and increase its spin rate before firing its own upper stage rocket forty-five minutes later. The satellite just remained dormant in orbit.

Ground crews frantically began trying to determine what went wrong. The consensus was that somehow a 4-inch switch that activated the satellite wasn't thrown—a 4-inch switch had crippled a 7.5 ton, $85 million satellite!

How could they throw the switch and activate the satellite? Sending an astronaut out to meet the Syncom was considered too dangerous because the satellite's rocket was still fueled. Finally the decision was made to try to tug the switch with jury-rigged "flyswatters" attached to the end of the robot arm. On the fifth day of the mission, astronauts Jeff Hoffman and David Griggs attached the homemade flyswatters, made of notebook covers, wire, and tape, in a three-hour EVA.

The next day *Discovery* went after the dead Syncom. With Rhea Seddon at the controls of the robot arm the crew pulled up alongside the satellite and Rhea successfully tugged the switch with the flyswatters. It was a bitter disappointment when the satellite refused to respond. Syncom's problem obviously lay elsewhere. Mission control congratulated the crew on their valiant effort and called off the hunt. There was nothing else the astronauts could do.

Despite Syncom's failure, the mission displayed the shuttle's ability to respond to the unexpected

—as shown in the first unplanned extravehicular activity. The shuttle proved its mettle once again.

After the mission ended NASA was already studying options for a repair attempt, and decided to conduct a repair mission on the stranded satellite during a future shuttle flight.

STS-24: Spacelab Returns to Orbit

Only ten days after *Discovery* landed, *Challenger* was poised on the launch pad with Spacelab in the cargo bay. Designated Spacelab-3 (Spacelab-2 had been delayed), this mission carried fifteen experiments covering four broad scientific disciplines: astrophysics, atmospheric observation, life sciences, and materials science.

These experiments returned over 250 million bits of data to earth, with some of the more spectacular returns coming from the atmospheric observation and materials science experiments. In the auroral-imaging experiment, Don Lind was able to photograph eighteen separate auroral displays resulting in the most detailed study of the phenomenon ever made from a manned spacecraft. The accompanying photograph of Aurora Australis captured over the south polar region is representative of the intense geomagnetic storms that occur when the atmosphere is disturbed by solar activity.

Two materials science experiments had very significant results with long-range applications. A mercuric-iodide crystal growth experiment resulted in a high quality crystal 300 to 400 percent larger than preflight predictions. Mercuric-iodide crystals have X-ray and gamma ray detection applications, and the larger, purer, space-grown crystals are much better than those produced on earth.

In another highly successful experiment, Taylor Wang of the Jet Propulsion Laboratory conducted a containerless processing technique that

Top: "Flyswatters" on the end of the arm approach the dormant Syncom.

Above: Aurora Australis as photographed by *Spacelab-3.*

Opposite: Taylor Wang conducting containerless processing experiment.

long term potential for space processing applications. Wang was able to levitate and manipulate a fluid using only ultrasonic waves. Many future space-processing experiments require containerless processing to produce unique glasses and metals (discussed further in chapter 8).

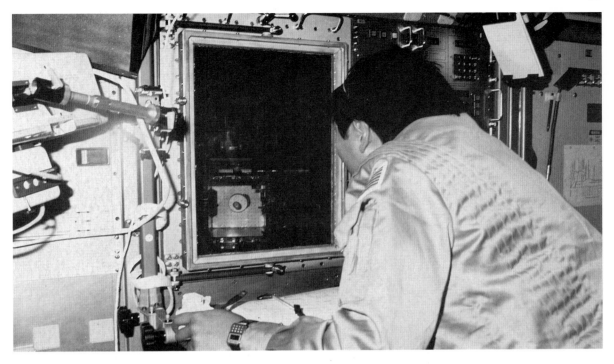

Shuttle Missions Into the 1990s

Because they are quite efficient at placing satellites into geosynchronous orbit, expendable launch vehicles will continue to play a role in launching satellites. In low earth orbit, however, the shuttle has unique advantages: the presence of astronauts; the ability to carry very large payloads such as Spacelab; the repair and retrieval capability; and the ability to return payloads to earth. These capabilities make the shuttle the cornerstone of future space activities.

In January 1986, *Challenger* (mission 51-L) was destroyed in an explosion 72 seconds after launch. The loss was a terrible blow to the shuttle program, but one that will only slow it down, not stop it. Additional orbiters will be built to carry out the space station program and proof-of-concept tests on the Strategic Defense Initiative components. With only a three-orbiter fleet, NASA will be hard put to meet its plans for the eighties, let alone the nineties.

Upcoming shuttle missions are still scheduled at a rapidly increasing rate, each one a stepping stone, providing experience and developing the foundation that will enable us to greatly expand the utilization of space. The shuttle flight manifest is practically booked solid through late 1989. Flight plans call for seventeen missions in 1987, and twenty-four missions annually in 1988 and 1989. Reservations have been made for over 190 major payloads. These payloads consist of planetary probes, astronomy experiments, military satellites, space laboratories, earth-sensing satellites, and new high-power communication satellites. Over 55 percent of these payloads would be impossible without the shuttle, because of their

Space Shuttle Missions

Mission/ Orbiter	Date	Crew	Mission Duration Days:Min:Sec	Earth Orbits	Miles Traveled	Payloads
STS-1 *Columbia*	Apr 12–14, 1981	John Young Bob Crippen	02:06:21	36	1,074,000	• Development Flight Instrumentation (DFI)
STS-2 *Columbia*	Nov 12–14, 1981	Joe Engle Dick Truly	02:06:24	36	1,074,000	• Office of Space & Terrestrial Applications (OSTA-1) • DFI
STS-3 *Columbia*	Mar 22–30, 1982	Jack Lousma Gordon Fullerton	08:00:06	130	3,800,000	• Office of Space Sciences (OSS-1) • DFI • Monodisperse Latex Spheres
STS-4 *Columbia*	Jun 27– Jul 4, 1982	Tom Mattingly Henry Hartsfield	07:01:10	112	3,300,000	• DOD 82-1 • DFI • Electrophoresis—McDonnell-Douglas-Johnson & Johnson • Monodisperse Latex Spheres
STS-5 *Columbia*	Nov 11–16, 1982	Vance Brand Bob Overmyer Bill Lenoir Joe Allen	05:02:15	81	2,118,000	• Satellite Business Systems-C • Telesat-E (Canada) • DFI
STS-6 *Challenger*	Apr 4–9, 1983	Paul Weitz Karol Bobko Story Musgrave Don Peterson	05:00:24	80	2,092,000	• Tracking & Data Relay Satellite • Electrophoresis—MDAC-J&J • Monodisperse Latex Spheres • First EVA
STS-7 *Challenger*	Jun 18–24, 1983	Bob Crippen Rick Hauck Sally Ride John Fabian Norman Thagard	06:02:25	98	2,530,000	• Telesat-F (Canada) • Palapa-Bl (Indonesia) • Shuttle Pallet Satellite (SPAS-01) • OSTA-2 (Materials Processing) • Electrophoresis—MDAC-J&J • Monodisperse Latex Spheres
STS-8 *Challenger*	Aug 30– Sep 5, 1983	Dick Truly Dan Brandenstein Guy Bluford Dale Gardner Bill Thornton	06:01:08	97	2,500,000	• First night launch-landing • Insat 1-B (India) • Remote Manipulator System-Payload Flight Test Article • Electrophoresis—MDAC-J&J
STS-9 (41-A)** *Columbia*	Nov 28– Dec 8, 1983	John Young Brewster Shaw Robert Parker Owen Garriott Byron Lichtenberg Ulf Merbold	10:07:36	166	4,600,000	• Spacelab-1 (77 Experiments) • Astronomy • Solar-Space Physics • Materials Science • Life Sciences • Atmospheric-Earth Observation
STS-11* (41-B)** *Challenger*	Feb 3–11, 1984	Vance Brand Bob Gibson Bruce McCandless Ron McNair Bob Steward	07:23:44	128	3,311,380	• First KSC Landing • Palapa B-2 (Indonesia) • Westar VI (Western Union) • SPAS-01a • EVA with Manned Maneuvering Unit

*Missions STS-10, and -12 were postponed due to problems in certifying the upper stage that would carry satellites on those missions.

**Beginning with the STS-9 mission, NASA adopted a new numbering system for payload assignments. The first digit represents the fiscal year; the second digit designates the launch site (1 = KSC, 2 = VAFB); and the letter is the alphabetical sequence of missions in that year.

Space Shuttle Missions

Mission/ Orbiter	Date	Crew	Mission Duration Days:Min:Sec	Earth Orbits	Miles Traveled	Payloads
STS-13*** (41-C)**** *Challenger*	Apr 6–13, 1984	Bob Crippen Dick Scobee George Nelson Terry Hart James van Hoften	06:23:40	107	2,880,000	• Solar Maximum Repair • Long Duration Exposure Facility (LDEF-1)
STS-14*** (41-D)**** *Discovery*	Aug 30–Sep 5, 1984	Henry Hartsfield Michael Coats Judy Resnik Dick Mullane Steve Hawley Charles Walker	06:00:57	96	2,498,755	• Syncom IV-2 (U.S. Navy) • Satellite Business Systems-D • Telstar 3-C (AT&T) • Solar Array Demonstration • Electrophoresis—MDAC-J&J
STS-17*** (41-G)**** *Challenger*	Oct 5–13, 1984	Bob Crippen Jon McBride Kathy Sullivan Sally Ride Dave Leestma Marc Garneau Paul Scully-Power	08:05:25	132	3,434,684	• Earth Radiation Budget Satellite (ERBS) • OSTA-3-Shuttle Imaging Radar-B • Large Format Camera • Orbiter Refueling Demonstration
STS-19*** (51-A)**** *Discovery*	Nov 8–16, 1984	Rick Hauck David Walker Anna Fisher Dale Gardner Joe Allen	07:23:46	126	3,289,406	• Palapa B-2 Retrieval • Westar VI Retrieval • Syncom IV-1 (U.S. Navy) • Telesat-H (Canada)
STS-20 (51-C)**** *Challenger*	Jan 24–27, 1985	Tom Mattingly Loren Shriver Ellison Onizuka James Buchli Gary Payton	03:01:34	47	1,229,542	• Department of Defense
STS-23*** (51-D)**** *Discovery*	Apr 12–19, 1985	Karol Bobko Donald Williams Rhea Seddon Jeffrey Hoffman David Griggs Charles Walker Senator Jake Garn	06:23:56	109	2,889,785	• Telesat-I (Canada) • Syncom IV-3 (U.S. Navy) • Electrophoresis—MDAC-J&J

STS-24
(51-B)****
Challenger
	Apr 29–May 6, 1985	Bob Overmyer
		Fred Gregory
		Don Lind
		Bill Thornton
		Norman Thagard
		Taylor Wang
		Lodweijk van den Berg
			07:00:08	110	2,900,000	• Spacelab 3 (2nd Spacelab mission)
						  • Life Sciences
						  • Materials Science
						  • Atmospheric Observation
						  • Astrophysics

***Missions STS-15, and -18 were postponed due to problems in certifying the upper stage that would carry satellites on those missions. STS-16 was combined with STS-14. STS-21 and -22 were delayed to later in 1985.

****Beginning with the STS-9 mission, NASA adopted a new numbering system for payload assignments. The first digit represents the fiscal year; the second digit designates the launch site (1 = KSC, 2 = VAFB); and the letter is the alphabetical sequence of missions in that year. This new numbering system was supposed to clear up the skipping of numbers, but as you can see the letters jumble just as easily when there are slips and delays.

70

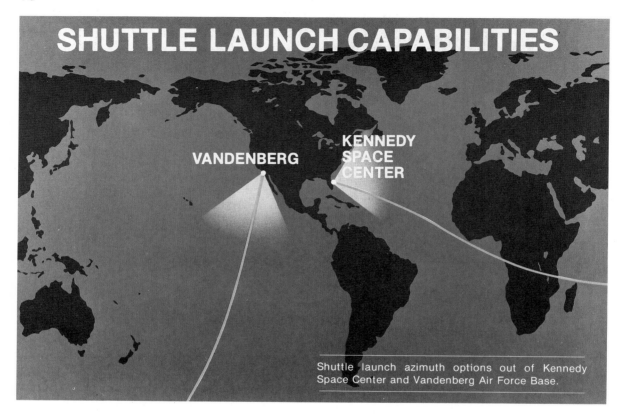

SHUTTLE LAUNCH CAPABILITIES

VANDENBERG

KENNEDY SPACE CENTER

Shuttle launch azimuth options out of Kennedy Space Center and Vandenberg Air Force Base.

size or operational requirements.

There are also over 450 small, self-contained payloads, dubbed "getaway specials," that will fly aboard the shuttle. These low-cost getaway specials ($3,000 to $10,000) provide access to the space environment for a wide variety of experimenters who have been excluded from the space program in the past because of the cost. Many getaway-special users take advantage of these low prices for conducting small-scale experiments to test new concepts prior to committing funds for full-scale projects. Shuttle missions will carry anywhere from one to twenty of these small payloads per mission depending on available space. Getaway specials have already been flown by such diverse users as high schools, universities, small businesses, foreign governments, and even private

citizens! Experiments to date have covered areas ranging from biology to materials science to astronomy. Basically anything that can fit in a getaway special canister will be allowed to fly as long as it doesn't pose a danger to the shuttle.

The Shuttle Fleet

Initially Congress authorized a fleet of four orbiters. With the loss of *Challenger,* the need for additional orbiters is under study and as demand continues to grow the current orbiters— *Columbia, Discovery,* and *Atlantis*—will be joined by their sister ships in the 1990s.

Although they appear identical, each new vehicle is more capable than the previous one.

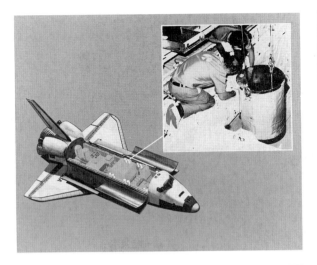

Installation of a typical "getaway special."

Through the use of weight-reduction and engine-performance improvements, each new orbiter has a greater launch capability than its predecessor. In addition, to give the shuttle increased flexibility, a second launch facility has been built at Vandenberg Air Force Base in California. Located about 120 miles north of Los Angeles, Vandenberg is scheduled to become operational in 1986.

The Vandenberg launch facility will enable the shuttle to be launched into orbit over the earth's poles. A satellite in polar orbit has the ability to view the entire earth spinning beneath it, as opposed to a satellite in equatorial orbit (in conjunction with the earth's rotation), that views only a portion of the earth. The launch facilities at Kennedy Space Center in Florida are limited to low inclination launches, either due east or northeast, but not north or south. Polar orbit missions are not allowed at Kennedy Space Center because they would require flying over populated areas.

Beyond 1989 projections show a sustained rate of twenty-four shuttle missions per year. However, if demand continues to grow and a fifth, or even a sixth orbiter is built, there could be as many as thirty to forty missions per year in the 1990s.

THE ORIGIN OF ORBITER NAMES

Because they are the ships that will explore and develop the ocean of space, it is appropriate that each of the first four orbiters were named after famous sailing ships:

Columbia was named after two famous ships: one of the first Navy ships (1836) to circumnavigate the globe; and the *Apollo 11* command module which carried Neil Armstrong, Michael Collins, and Buzz Aldrin on the first lunar landing mission, July 16–24, 1969.

Challenger was also named after a Navy ship: the original ship *Challenger* made an extended exploration of the Atlantic and Pacific Oceans from 1872 to 1876 and the name was also used for the *Apollo 17* lunar module.

Discovery was named after two famous ships: the one in which Henry Hudson discovered Hudson Bay while searching for the northwest passage between 1610 and 1611; and the ship on which Captain Cook discovered the Hawaiian Islands and explored southern Alaska and western Canada.

Atlantis was named after a two-masted ship operated for the Woods Hole Oceanographic Institute from 1930 to 1936. It traveled more than a half million miles conducting oceanic research.

The earth as seen from the moon—notice the western United States and Baja California.

CHAPTER 6

WHY SPACE?

"WE DO NOT REALIZE WHAT WE HAVE ON EARTH UNTIL WE LEAVE IT."

JAMES M. LOVELL
APOLLO 8 ASTRONAUT

Many of the significant lessons we have learned from space exploration have been about our own planet. Early photographs of the earth as seen from the moon stunned the world. From tens of thousands of miles away the earth appears quite small, a beautiful marble colored in blues, greens, browns, and oranges rotating against the starry black vastness of the cosmos. This new perspective has given us a better appreciation of the earth. From space, the earth can be seen as the enormous ecological system it is—exceedingly complex, with infinite varieties of interrelated and interdependent parts. The quality of life here—in fact, our very survival on this planet—depends on the preservation of that system. In order to preserve it we must find ways to use it more efficiently.

Space gives us many unique methods to do just that. Space exploration has spawned the development of new satellite technologies that enhance life on earth. Among these are communication satellites, which are in an explosive growth era today; meteorological satellites, an invaluable aid in weather prediction; earth observation satellites, used to track air and sea pollution as well as aiding in the location of natural resources; navigation satellites, for greater accuracy; and spy satellites, without which arms control agreements would be virtually impossible. With the expanded use of the space shuttle, space activities will continue to grow at a rapidly accelerating rate during the coming years.

Communication Satellites

It wasn't long ago that the world marveled at early broadcasts from communication satellites —instantaneous news flashes and coverage of the 1964 Olympic games from Japan that proclaimed, "live via satellite." In today's world that phrase has become so commonplace that it is seldom used. Because of satellites, it is now possible for more than a billion people to witness an event as it happens.

Significantly, communication satellites have

greatly increased the telephone's impact on society. For instance, nearly 250 countries can be reached from almost any telephone in America, and these calls cost less than ever. In 1965 at the beginning of the communication satellite revolution, the cost of a three-minute phone call from the United States to London was $9.00. Since that time the price has dropped steadily. Today that same phone call is only $3.00. Moreover, if the price of that phone call had grown with the rate of inflation (as have most products), today's price would be more than $27.00! Since it did not, the number of transoceanic phone calls have grown from 3 million in 1965 to more than 200 million per year in the early 1980s. Communication satellites have truly brought the people of the world closer together.

This dramatic increase was possible because communication satellites have grown larger and are capable of handling greater amounts of traffic at reduced cost. For example, *Early Bird,* one of the first commercial communication satellites, weighed 150 pounds and had only 240 voice circuits. Today a typical communication satellite weighs 3,000 to 4,000 pounds and has 10,000 to 15,000 voice circuits. As the space shuttle era takes

Above: Satellite Business Systems satellite prior to launch by shuttle, 1982.
Below: Technician servicing *Early Bird* prior to launch, 1965—"Live via satellite."

hold, larger satellites will make today's satellites appear as simple as the *Early Bird.*

In today's information-based society and with projected increases in world population, the need for education, information, and communications is ever increasing. Space has overwhelming advantages in providing these services. A communications satellite in a 22,300-mile geosynchronous orbit is accessible to an entire nation for communications, educational TV, and information exchange. (Although we look upon television primarily as entertainment, it has the potential to be one of our most powerful educational tools.) While building their national communications networks, developing nations have the opportunity to leapfrog a generation of technological advancement by using communication satellites, the use of

A Canadian communication satellite's "footprint."

which eliminates much of the costly and time-consuming process of laying cable across their nations. Mexico, Brazil, Colombia, India, Indonesia, and the Arab countries are all building satellite networks that will greatly aid their economic advancement.

As satellites become larger to take advantage of the shuttle's ample cargo bay, they will use higher frequencies, which will in turn allow the use of smaller, less complex receiving antennas. Earth antennas will shrink from the 10-foot antenna typical of the 1970s and early 1980s, to three-foot rooftop antennas in the latter part of the decade. The reduced cost and ease of operation of these smaller antennas is already expanding the use of communication satellites.

This incredible "shrinking act" will accelerate when satellite builders begin taking advantage of the shuttle's cargo capability. The shuttle's cargo capacity is essentially unlimited because astronauts will be able to construct enormous satellites carried as separate components by multiple shuttle missions. Satellite receivers could shrink to wristwatch size!

Video Teleconferencing

One promising new communication-satellite service currently in its infancy is video teleconferencing. To conduct a typical teleconference, large TV screens are assembled at numerous locations and the meeting is conducted by satellite without the expensive and time-consuming chore

of traveling. Thousands of people can meet simultaneously worldwide.

First introduced in 1977, teleconferencing has grown to over 100,000 meetings per year and will continue to grow rapidly as advanced equipment becomes available. Many firms are reducing travel costs by using teleconferencing to conduct meetings with their widely dispersed sales forces rather than meeting at a central location. Large hotel chains, such as Hilton and Holiday Inn, now have permanent teleconferencing equipment in many of their hotels. By reducing the need for travel, teleconferencing also makes a major contribution to energy conservation.

Search and Rescue Satellite System (SARSAT)

Another new communication-satellite service is the multinational search and rescue satellite system (SARSAT). SARSAT is a joint project of the United States, Canada, France, and the Soviet Union, which began as an experimental program in September 1982.

The system consists of electronic packages that

METEOROLOGY

SYNCHRONOUS

LOW EARTH ORBIT

Opposite: Future satellites will shrink receivers to wristwatch size.

Above: Weather satellites, the big picture and close-up.

piggyback aboard U.S. and Soviet weather satellites and listen continuously on emergency frequencies used by ships and aircraft. When the satellite detects an alert it relays the information to ground stations. Within minutes the station's computer produces a position fix, locating the distressed craft within a radius of 3 to 12 miles.

With such accurate positioning, rescue crews can locate the troubled party in a fraction of the time required by conventional methods—greatly increasing the likelihood of survivors. In its first three years of operation, SARSAT was credited with saving the lives of over 350 people in downed aircraft and maritime accidents. The program's wide success has led the four SARSAT nations to sign an agreement continuing the program well into the 1990s.

Weather Satellites

What modern-day TV weatherman could be without a weather satellite photo during his daily forecast? While weather satellites are often taken

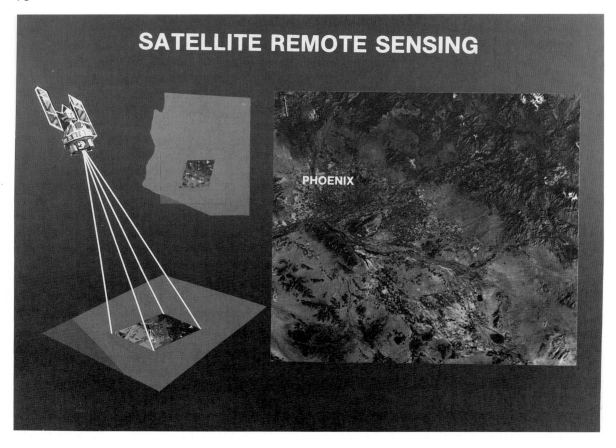

SATELLITE REMOTE SENSING

PHOENIX

for granted, they provide data that is available by no other means.

Before weather satellites came on the scene in 1960, our knowledge of global weather conditions was sketchy at best. Hot air balloons and airplanes were used to track weather patterns, and two-to-three-day weather forecasts were still a vision of the future. Today, sophisticated satellites using infrared sensors can observe developing weather patterns day and night.

The U.S. weather satellite system consists of two types of weather satellites: those in geosynchronous orbit (22,300 miles above the equator) that give the big picture for study of global weather patterns, and others in lower orbits (300

Typical 115-mile by 115-mile Landsat photo.

miles) that provide detailed information on regional conditions. In addition to providing everyday forecasts, weather satellites give vital information to everyone from fishermen and sailors to farmers and airline personnel. By providing early warning of hurricanes and other severe storms, giving frost warnings, and measuring wave heights, weather satellites have saved countless lives and millions of dollars over the years.

The combination of increasingly sophisticated weather satellites and an extensive network of ground systems, all tied together by powerful computers, is enhancing the ability to conduct long-term weather forecasting. As we continue to understand global weather patterns, we may some day have the capability to alter them. As an example, the location and tracking of clouds ripe for seeding could produce rain in regions suffering from severe drought. On the other hand, hurricanes could be weakened by seeding them before they gather too much strength.

Earth Observation Satellites

Working much like weather satellites, earth observation satellites are also circling the globe. The difference is that earth observation satellites use sensors operating in various bands of the electromagnetic spectrum, ranging from the visible to radar and the infrared range. This type of sensing is possible because everything—living or non-living—emits, absorbs, and reflects electromagnetic radiation in its own distinctive way. Minerals, vegetation, water, pollution, all radiate unique signals that can be picked up by satellite.

The most widely used earth observation satellite is the Landsat system, the first of which was launched in 1972. Landsat satellites take photos, each covering an area 115 miles by 115 miles; this enables surveillance of large and hard-to-reach areas. Most significantly, these surveys can be performed economically in minutes, instead of the hours, days, or months required for ground or aircraft surveys.

Data from Landsats and other earth observation satellites supply invaluable information that enables us to determine what is happening in our world today, as well as how it changes over time. Government and the private sector use Landsat photos to forecast crop yields, and detect disease and insect infestation of crops. In the United States alone, forty states use Landsat data for their natural resource planning including air and water pollution assessment, forest inventory, water and geological mapping, and wildlife habitat inventory.

Earth Observation and Shuttle Missions

Future earth observation satellites are being developed with the aid of space shuttle missions. The second shuttle flight (STS-2, November 1981) demonstrated that space components can be tested aboard the shuttle and returned to earth for improvement and subsequent reuse in space.

The synthetic aperture radar (SAR) remained in the STS-2 cargo bay throughout the mission. The SAR not only imaged the surface of the earth as planned, but also provided imagery of subsurface features, revealing an ancient network of rivers buried under the sands of the Sahara desert. Radar waves are able to penetrate dry soil and in the Sahara, the most arid region of the world, the SAR was able to penetrate the sand as deep as 6 to 20 feet. Since this discovery it has been determined that there were flowing rivers in the area during the time of the pyramids, nearly 5,000 years ago.

Such subsurface-imaging data has broad implications for archaeological studies as well as for oil and water exploration in similarly dry areas where radar could penetrate the surface down to bedrock level. For archaeologists the radar could be used to help determine sites of early human habitation near former rivers and lakes. For geologists it may reveal subsurface features similar to currently successful oil and mineral locations.

The synthetic aperture radar was further improved, flown again on STS-17 in October 1984, and will continue to fly on the shuttle on a periodic basis. Lessons learned on these missions will aid the development of advanced radar satellites.

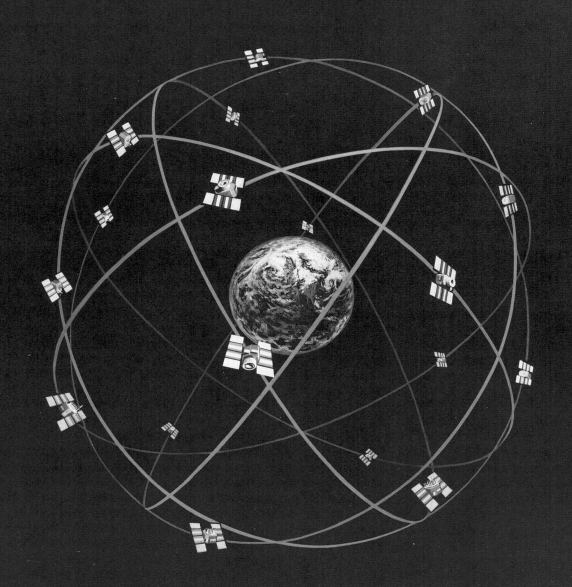

CHAPTER 7

THE MILITARY USE OF SPACE

"OUR PERCEPTION OF SPACE HAS CHANGED. IT IS NOW SEEN THAT SPACE IS A PLACE—LIKE THE LAND, AND SEA AND AIR—ANOTHER DIMENSION. AND, IT WAS JUST A MATTER OF TIME UNTIL WE STARTED TREATING IT AS SUCH. OUR VIEW IS THAT THIS MOVE (CREATION OF THE SPACE COMMAND) WILL SUBSTANTIALLY CONTRIBUTE TO DETERRENCE, AND THUS INTERNATIONAL STABILITY, BY REDUCING UNCERTAINTY."

GENERAL JAMES V. HARTINGER,
FORMER COMMANDER,
U.S. SPACE COMMAND
JULY 1983

The military uses space much the way civilians do—gathering and sending information by satellite. Reliable communications, better weather information, and accurate navigation are vital to a commander. The high vantage point of space provides the most efficient way to accomplish these functions, and more significantly, space provides

---

NAVSTAR—a new dimension to navigation.

a calming effect to the world. Surveillance satellites help keep the peace by reducing the chances of a surprise attack.

Surveillance satellites provide knowledge of large military movement anywhere in the world. Mass troop concentrations can no longer be conducted without detection. Just consider the difference this would have meant in World War I or World War II. Furthermore, surveillance satellites enable world leaders to sign arms agreements. Without these satellites we would probably still be in a state of anxiety similar to that of the cold war of the 1950s.

The military's use of space is a logical extension of what has been the basic objective of armies for centuries: to gain the high ground for enemy observation, and to communicate with and control your own forces. Space is the ultimate high ground; by using satellites, armed forces on land, at sea, or in the air become more effective. Satellites are often referred to as force multipliers because of their ability to enhance the effectiveness of existing forces.

Reliable communications are at the heart of all military operations. Some of the most devastating defeats in military history have proven time and again that if one cannot communicate reliably,

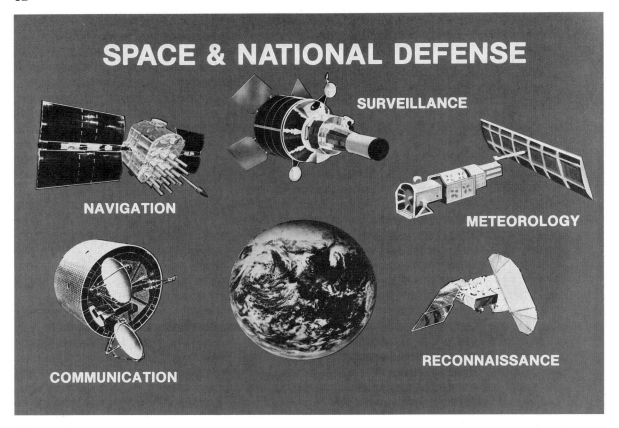

SPACE & NATIONAL DEFENSE

SURVEILLANCE

NAVIGATION

METEOROLOGY

COMMUNICATION

RECONNAISSANCE

one cannot control a force. Troops can be mispositioned, get lost, or lose track of the enemy. Communication satellites provide a reliable link that can help control forces; the Pentagon today relies on satellites for over 70 percent of long-haul communications.

Navigation via the Global Positioning System

Prior to the 1950s, navigation was still largely dependent on celestial readings which provided

Artist's concept of various military satellites and their functions.

accuracy within 10 to 15 miles. This was enhanced by inertial navigation which gave accuracies within 1 or 2 miles. In the 1960s, the Navy began experimenting with satellites for navigation and by the 1970s they were providing accuracies of approximately 1,500 feet. This was a great step forward, but a position could still be off by the size of five football fields—definitely not good enough for a large aircraft carrier or a jet travelling at Mach 2!

One of the most exciting and promising developments in the military use of space is the Navstar global positioning system (GPS), a satellite navigation system. When completed, the GPS will consist of a "constellation" of eighteen satellites located approximately 11,000 miles above the earth.

By using the GPS, ground forces, ships, and aircraft will be able to determine their location with unprecedented accuracy. Receivers will translate satellite signals into navigational information enabling the user to know his position with 30 to 50 feet accuracy anywhere in the world, at any time of day, and in any kind of weather. The receivers also can provide accurate altitude and speed data. Best of all it is a passive system: the satellites send continous signals worldwide that users can pick up without being detected.

GPS applications currently being tested or under consideration include use by strategic and tactical aircraft, helicopters, submarines, aircraft carriers, infantry vehicles and armored vehicles. Test results have already revealed the following:

- Helicopters can make blind landings at night within several feet of a designated spot

- Cargo aircraft have parachuted supplies within 30 to 40 feet of a ground marker

- Blind bombing runs at night with conventional (unguided) bombs have come within 10 to 20 feet of the target center

- Large ships have been safely guided out of unfamiliar harbors in low visibility.

These excellent results were accomplished using partial constellations of six to eight experimental satellites. The shuttle will begin launching global positioning system satellites in 1986, and the full constellation of eighteen satellites is scheduled to become operational by 1988.

Possible civilian use of the system is also being considered for improved air traffic management, shipping navigation, and even use in the family car. Ford, Chrysler, and General Motors have already produced receiver prototypes that display a computerized map on a video screen. The screen fits in the car's console and shows a symbol of your car as it progresses along. These auto navigational units should be available for about $500 in the late 1980s when the system becomes fully operational.

The potential benefits of civilian GPS use was tragically pointed out in September 1983 when the Soviet Union attacked an off-course Korean Airline 747 jumbo jet killing all 269 passengers. Following the incident, Senator Charles Percy, chairman of the Senate Foreign Relations Committee, called for civilian use of Navstar. Although the airliner's erratic course was attributed largely to crew error, if that Korean airliner had been using the global positioning system it would have had a better chance of avoiding Soviet airspace and 269 lives could have been saved.

Soviet Military Activities in Space

Beaten in the race to the moon, the Soviets have concentrated their efforts on using space to aid their military. The accompanying chart shows the tenacious Soviet pursuit of space domination. During the 1970s, the Soviets launched almost *four times* more missions than the United States, averaging over ninety launches per year, and this trend has been accelerating.

In 1982 the Soviets established an all-time record of launching one hundred and one missions. Not even a major explosion at one of their launch pads in 1983 slowed the Russian space program—they still launched ninety-eight missions that year and ninety-seven in 1984. By comparison, the United States launched only twenty-one missions in 1984. While these numbers are somewhat misleading (most U.S. satellites are more sophisticated and last longer), the sheer volume of Soviet launches is impressive. The trend is unmistakable.

Although the space shuttle will help close the

Scale-model drawing of the Soviet *Salyut-6* space station.

gap in the 1980s, the percentage of Soviet launches dedicated to military activity—80 percent military versus 25 percent for the United States—shows the fundamental difference in philosophy between the two nations.

The Soviets have sought to establish a *permanently* manned presence in space. They have launched seven increasingly sophisticated space stations beginning in 1971 with their first station, *Salyut 1,* through 1982, with the launch of *Salyut 7.* While some of the Salyut stations have been scientifically oriented, others have reportedly been primarily military.

The accompanying illustration shows the *Salyut 6* space station. Though designed to last three years, *Salyut 6* functioned for four and a half years before being replaced by *Salyut 7* in 1982. *Salyut 6* was visited by astronauts from Cuba, Vietnam, Czechoslovakia, Poland, Bulgaria, Hungary, and East Germany in addition to *fourteen* Soviet crews.

The Russian goal of a permanently manned space station is practically reality. In 1984 a two-man crew remained aboard *Salyut 7* for 238 days —almost eight months! Prior to that record-shattering mission, a 211-day mission was conducted aboard *Salyut 7* in 1982, and two long-duration missions of 175 and 185 days were conducted aboard *Salyut 6* in 1979 and 1980 respectively.

Soviet cosmonauts have spent five times as many hours in space as U.S. astronauts. Soon the Soviets will launch space stations that could accommodate permanent crews of twelve, and future plans call for even larger space stations with crews of thirty to fifty.

By contrast, the only U.S. space station to date has been the Skylab program of 1973 to 1974. Although studies are underway to develop a new U.S. space station, the earliest it could be launched would be the early to mid-1990s. Furthermore, U.S. space station studies have concentrated on civilian activities.

The Militarization of Space

The launching of *Sputnik* in 1957 signaled a new era and a new arena of competition between the superpowers. Both the United States and the Soviet Union began a race to see which country could establish its preeminence. The United States, with its spectacular moon landings and its much-publicized space programs, has certainly won the publicity battle. But the Russians, with their persistent military effort in space, may be winning a growing competition to develop new weapons for use in outer space.

The Soviet military space program has grown to unprecedented levels. And the Soviets use space offensively as well as defensively. They have tested hunter-killer satellites (more commonly known as antisatellites or ASATS) since 1968. Since that time there have been twenty ASATS tested, half of which have reportedly been successful.

The Soviet ASAT has the effect of a grenade in space. It moves alongside its target satellite and explodes, showering the surrounding area with shrapnel. Theoretically, one piece of shrapnel can knock out a satellite by damaging its sensitive electronics or its power-producing solar cells. In a major war these Soviet ASATs would undoubtedly be used to knock out U.S. surveillance satellites, in effect "knocking our eyes out."

The United States does not yet have anti-satellites. A U.S. ASAT is under development and was tested in 1984 and 1985, but is not expected to become operational until the late 1980s. The Soviets have called for a ban on all ASAT testing and development. Considering the fact that Russia already has an operational ASAT, it is not a surprising request!

In response to this vigorous Soviet space activity and recognizing the increasing reliance on space systems to support ground forces, the United States established the Air Force Space Command in September 1982. By establishing a formal command structure, the Air Force has given space activities the same priority ranking as other major Air Force Commands such as the Strategic Air Command or the Military Airlift Command.

Headquartered in Colorado Springs, Colorado, the Space Command operates four bases and twenty-two units located around the globe. Its functions include control of all operational Air Force satellite systems (communications, weather, global positioning, etc.), control of ground-based missile warning units throughout the world, and

Relentless Soviet pursuit of space leadership.

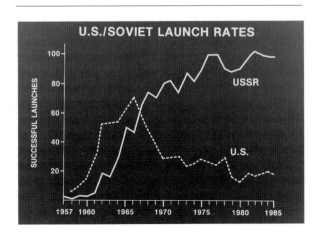

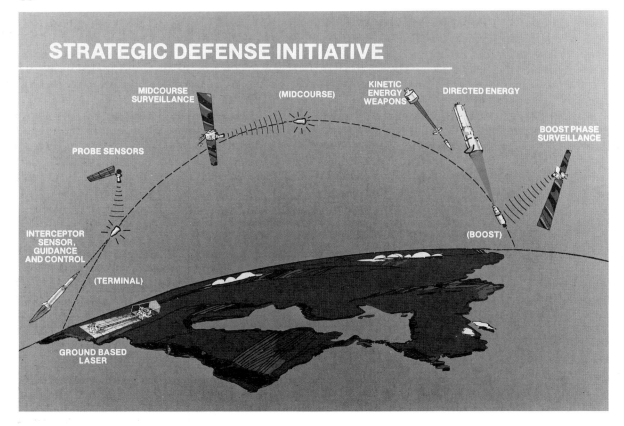

STRATEGIC DEFENSE INITIATIVE

MIDCOURSE SURVEILLANCE

(MIDCOURSE)

KINETIC ENERGY WEAPONS

DIRECTED ENERGY

PROBE SENSORS

BOOST PHASE SURVEILLANCE

INTERCEPTOR SENSOR, GUIDANCE AND CONTROL

(BOOST)

(TERMINAL)

GROUND BASED LASER

A "layered defense."

responsibility for all military-oriented space shuttle flights (approximately one-third of planned space shuttle flights). Space Command will also control the U.S. ASAT when it becomes operational, and space-based defense systems if the United States develops them.

A further step was taken by the Pentagon when it established the U.S. Space Command in September 1985. This unified command will include air force, navy, and army personnel, and will centralize responsibility for all military space systems. The new command was formed to make more efficient use of current space systems and to improve planning for future space operations.

Strategic Defense from Space — Opportunity or Folly?

During a March 23, 1983 address, President Reagan made the surprise commitment to undertake a research program focusing on a space-based strategic missile defense. Space-based directed energy (lasers or particle beams) and kinetic energy weapons could render intercontinental ballistic missiles (ICBMs) obsolete, thereby moving away from the Mutually Assured Destruction (MAD) doctrine that now exists.

Under the MAD doctrine it is assumed that neither side would begin a nuclear war because it would surely be destroyed by a retaliatory attack. Philosophically, the president's Strategic Defense Initiative is a major departure from the terror of the MAD doctrine. As the president commented on the research program, "Would it not be better to save lives than to avenge them?"

This type of defense has a certain appeal to it. Theoretically, since laser beams travel at the speed of light, about 186,000 miles per second, orbiting laser weapons could knock out Soviet nuclear missiles soon after they leave their silos. A typical ICBM would seem to be standing still as it travels between 12,000 and 15,000 miles per hour.

However, such a defense could also have a destabilizing effect depending on which nation developed it first. The Soviet Union reportedly is spending three to five times as much money as the United States to develop a space-based defense system. The nation that could first perfect such a system would have the option to launch a nuclear attack without fear of retribution, a very scary scenario.

At this time it's too early to tell exactly what a space-based strategic defense would look like, or whether one will be built. Such a system would basically be a "layered" defense along the lines of the accompanying drawing. It would probably include space weapons, surveillance satellites, and aircraft interceptors. Directed energy and kinetic energy weapons in space would attempt to intercept as many enemy ballistic missiles as possible in their boost phase when they are most vulnerable. Those that got through this first layer would be attacked by additional layers in mid-course, or by other layers during atmospheric reentry. Early studies show that up to seven layers might be required, involving hundreds of weapons. Such a system would be supported by an elaborate space surveillance system to locate and track the missiles. Furthermore, the space weapons might also be supplemented by ground-based directed-energy weapons that would use mirrors in space to direct their fire.

That is the theory, but many difficult obstacles must be overcome before this type of defense is possible. For instance, laser beams tend to dissipate as they travel through the air and enormous power supplies would be required to generate enough energy to penetrate a nuclear missile's shell. Furthermore, these weapons would have to be packaged small enough to be launched into space. Vehicles even larger than the shuttle might be required.

The Strategic Defense Initiative's most difficult task will probably be developing the enormous artificial intelligence required to simultaneously track, target, and destroy thousands of approaching missiles and decoys. Not only would the laser weapons have to find the missile and fire at it, they would also have to lock on to it for several seconds to be sure it was destroyed before moving on to the next missile.

These many obstacles have led many to call the Strategic Defense Initiative the president's "Star Wars" defense. Even the president alluded to these difficulties, stating that he didn't know when such a defensive system could be in place. Defense analysts predict that such weapons are ten to twenty years away. Many critics believe an effective strategic defense is impossible. At this point it is a research effort intended to give future presidents the option of expanding it to operational scale.

Whether effective space weapons are possible or not, the use of space for national security will continue to grow. The space shuttle will allow the U.S. military to carry out present missions with greater flexibility. And the Soviet buildup shows no sign of slowing. The upcoming years could be a major turning point for space activities. Space could remain a sanctuary for exploration, science, and commercial use. Or, space could become a potential battleground as military space systems change from a traditional support role into an operational role.

"Penguin" crystal grown aboard Skylab.

CHAPTER 8

MANUFACTURING IN SPACE

"ELEVEN SUCCESSFUL FLIGHTS OF THE SHUTTLE MEAN THAT WE ARE ON THE VERGE OF A SPACE TRANSPORTATION SYSTEM THAT CAN DEPENDABLY SUPPORT SPACE INDUSTRIES. AND THE BENEFITS OUR PEOPLE CAN RECEIVE FROM THE COMMERCIAL USE OF SPACE LITERALLY DAZZLE THE IMAGINATION."

PRESIDENT RONALD REAGAN ANNOUNCING THE NATIONAL SPACE COMMERCIALIZATION POLICY JULY 20, 1984

Manufacturing in space presents one of the most promising long-range prospects for the shuttle and for the space program as a whole. The shuttle's ability to carry large quantities of raw materials into space and return finished products to earth could spawn entirely new industries.

On earth we accept the pull of gravity as essential to our existence. In space things are radically different. For example, the Skylab astronauts discovered that without the compressing effects of gravity they were taller—up to three inches taller! Their spines had elongated to their full potential. After returning to earth (and gravity), they soon returned to their former sizes.

This is an extreme example of what gravity, or rather the lack of it, can do. Taken down to the molecular level, gravity seriously hampers many manufacturing processes on earth. The effects of gravity are most obvious in the case of sedimentation, the layering of materials of various densities. This is easily seen when trying to mix oil and water; water, being denser and heavier than oil, sinks to the bottom. Processing experiments have shown that stronger metal alloys can be produced

Zero gravity eliminates convection.

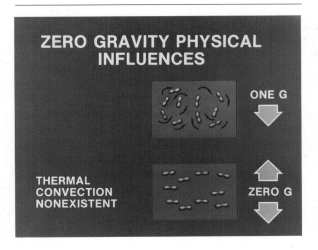

ZERO GRAVITY PHYSICAL INFLUENCES

ONE G

THERMAL CONVECTION NONEXISTENT

ZERO G

in space because the distribution of fibers during the melting and solidification stage is uniform; there is no sedimentation. Shuttle experiments have included such exotic alloys as silver-germanium, zinc-lead, copper-aluminum, and zinc-aluminum. These exhibit a strength-to-weight ratio far exceeding those of earth-produced alloys.

Thermal convection is another major inhibitor to producing certain products on earth. Thermal convection is best seen when boiling water: as water molecules are heated they rise to the surface as bubbles. In zero gravity convection is non-existent. And, with no sedimentation or convection to interfere with the process, materials of different densities can be melted, mixed, and resolidified to produce superstrong, lightweight alloys that are impossible to produce on earth.

One method of space manufacturing, containerless processing, was proven feasible during the Skylab program. Containerless processing is the ability to melt, solidify, or otherwise process a sample without physical contact. In zero gravity, a small acoustic or electromagnetic field is sufficient to hold materials stable as they are processed. This process eliminates impurities from containers which is a major problem in manufacturing certain materials (mostly metals and glasses) that have high temperature melting points. The containerless method will allow production of ultra-pure glasses for optical uses and new superstrong metal alloys.

The First Space Product

Latex spheres of microscopic size were the first space product to be sold commercially. Produced aboard the shuttle during four early missions, these tiny latex microspheres, which measure 10 microns (4/10,000ths of an inch), may have major medical and industrial applications for fields that demand precise measurement and calibration. Formally known as monodisperse latex spheres because of their uniform consistency, the spheres are perfectly round and exactly the same size. Their uniform consistency cannot be duplicated on earth due to gravity's influence.

These latex spheres are used as a kind of microscopic yardstick. The National Bureau of Standards now sells these tiny spheres as the standard for calibrating scientific equipment such as electron microscopes. The spheres are sold for $400 per finger-sized vial, each of which contains about 15 million spheres.

Future shuttle missions will attempt to produce larger spheres, ranging in size from 30 to 100 microns (40/10,000ths of an inch) which may have promising medical applications. For example, spheres could be used to measure the pores of malignant tumors and then be sized to fit inside the tumorous but not the healthy tissues. The spheres could then be filled with anticancer drugs to carry a more concentrated dose directly to the malignant tissue.

CONTAINERLESS PROCESSING

- FREE FLOATING MATERIALS SHAPED BY FORCE FIELDS
- ELIMINATES IMPURITIES & STRIATIONS
- ENHANCES STRENGTH & MECHANICAL PROPERTIES

Left: Advantages of containerless processing.

Right: Highly successful electrophoresis unit that flies in the shuttle middeck (lower crew compartment).

COMMERCIAL ELECTROPHORESIS

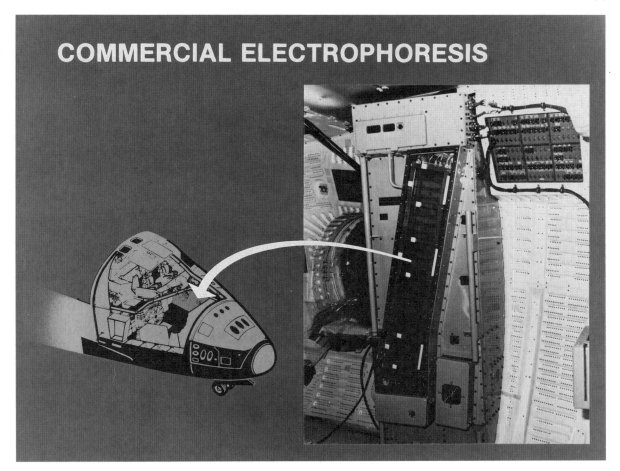

The Potential for Medical Breakthroughs

Other experiments carried aboard early shuttle missions have opened the potential for large-scale pharmaceutical processing in space. The most successful to date has been a joint venture between McDonnell Douglas and the Ortho pharmaceutical division of Johnson & Johnson that uses electrophoresis—the separation of cells by electricity—to separate biological materials.

Although used in laboratory research worldwide, gravity severely inhibits the effectiveness of electrophoresis thereby limiting its output to very small quantities.

During the shuttle experiments, streams of biological cells from the pancreas, kidneys, and pituitary glands were injected into a solution flowing through a six-foot rectangular chamber in the shuttle's middeck. When an electrical field was applied, the cells separated into individual streams that were collected as they flowed out the top of the chamber. In the zero-gravity environment of space the shuttle experiments separated *over 700*

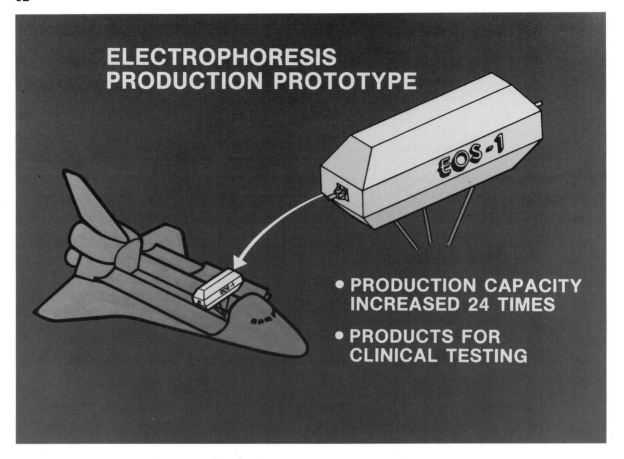

ELECTROPHORESIS PRODUCTION PROTOTYPE

EOS-1

● PRODUCTION CAPACITY INCREASED 24 TIMES

● PRODUCTS FOR CLINICAL TESTING

times more usable product *at five times* the purity than a similar experiment on earth. At that rate enough usable cells can be produced to begin clinical testing on humans.

Nearly fifty products have been identified for possible space manufacturing which could lead to dramatic breakthroughs in the treatment of anemia, cancer, diabetes, emphysema, dwarfism, thrombosis, and viral infection. Three of the most promising products being explored are beta cells, erythropoietin, and interferon.

Beta cells are the cells in the pancreas that produce insulin. Theoretically, large quantities of beta cells could provide a cure for diabetes, there-

by eliminating the need for daily insulin injection. But in the presence of gravity, it is not possible to isolate beta cells in quantities sufficient to treat patients. Transplants of beta cells have successfully controlled diabetes in laboratory animals and research is now expanding to human testing. The hope is that production in space will result in large enough quantities to treat diabetics on a wide scale.

Erythropoietin is a hormone produced by the kidneys which helps control red blood cell production in the body. The space-produced erythropoietin could be used to treat anemia sufferers, of which there are estimated to be approximately two

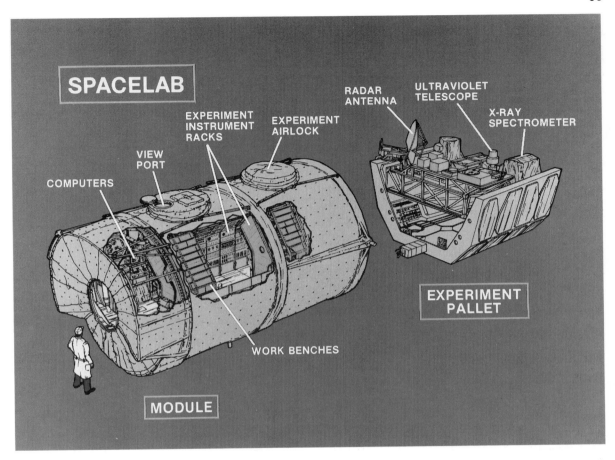

Opposite: Five-thousand-pound electrophoresis facility will fly in the cargo bay.

Above: Spacelab-1, a closer look.

and one-half million in the U.S. alone. A large percentage of these anemic individuals would benefit from erythropoietin treatment.

Interferon is a protein produced by the body as an immunological agent. It has been widely acclaimed as a possible cure for cancer and viral infections. However, earth-based processing has produced only small amounts of a low-purity product. Space-based processing promises the potential of large quantities of purer interferon.

Based on the success of the early shuttle experiments, McDonnell Douglas and Ortho have built a 5,000-pound prototype production facility designed to fly in the shuttle's cargo bay beginning in 1986. This unit, a possible forerunner of the commercial space factory, has twenty-four different separation chambers (there's only one chamber in the current device), and will be used to produce the large quantities of materials needed for human testing.

These steps are expected to lead to large-scale production aboard orbiting satellites that will be launched by the shuttle in six-month intervals beginning in 1988. The shuttle will then routinely

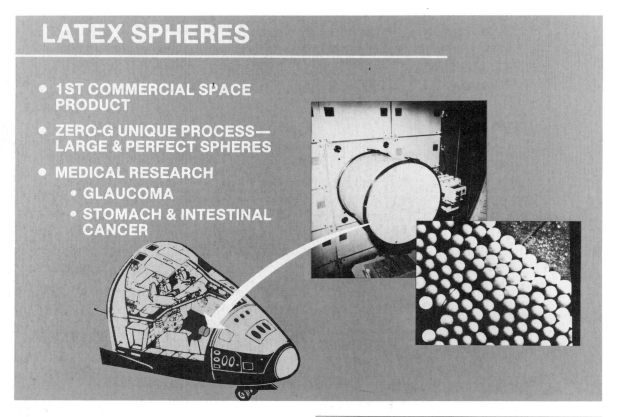

LATEX SPHERES

- **1ST COMMERCIAL SPACE PRODUCT**
- **ZERO-G UNIQUE PROCESS— LARGE & PERFECT SPHERES**
- **MEDICAL RESEARCH**
  - **GLAUCOMA**
  - **STOMACH & INTESTINAL CANCER**

visit these satellites every six months to supply additional raw material and to pick up the finished product for return to earth for commercial distribution. Even larger quantities could be produced aboard a permanent space station that will be in orbit in the 1990s.

This will be only the beginning of many space-based products. A growing number of firms are now experimenting with other processes that will benefit from space processing. Products under study include larger and purer crystals that could be used to produce semiconductors for the next generation of super-computers, ultra-pure glasses for telescope lenses and other optical uses, and unique metallic alloys for all kinds of products. The genesis of more space partnerships is taking shape today in the form of hundreds of space

Above: Shuttle-grown latex spheres magnified 1,000 times.

Opposite: Astronauts Owen Garriott *(foreground)* and Ulf Merbold in the Spacelab module during its first mission. Merbold, a West German, was the first European to fly aboard the shuttle.

manufacturing experiments planned for Spacelab missions.

Spacelab: A Laboratory in Space

Spacelab is a versatile, general-purpose laboratory that is carried in the shuttle's cargo bay. With the shuttle-Spacelab combination, scientific re-

searchers will no longer be earthbound. They will be able to go into space to perform research that cannot be done on earth.

The Spacelab program is a cooperative venture between NASA and the European Space Agency (ESA), a consortium of eleven European nations. ESA was responsible for the funding, developing, and building of Spacelab; NASA is responsible for its launch and operation.

Spacelab was developed on a modular basis and can be varied to meet specific mission requirements. Its two principal components are the cylindrical module, which is pressurized to provide a shirtsleeve working environment; and U-shaped pallets that directly expose telescopes, antennas, and sensors to space. With the shirtsleeve environment of the Spacelab module, scientists will enjoy many of the comforts of a ground-based laboratory, including computers, work benches, and instrument racks. Spacelab can be equipped with several tons of laboratory equipment for studies in all fields.

In November 1983, the first Spacelab mission, STS-9, showed off the lab's versatility. Scientists from Europe, Japan, Canada, and the United States provided seventy-seven experiments covering five research areas—astronomy, space plasma physics, atmospheric physics, life sciences, and material sciences—thirty-six of which were devoted to materials-processing research. They included tests on producing unique glasses, metal alloys, biologicals, and semiconductor materials that cannot be produced on earth.

Spacelab-1 was also the longest shuttle mission to date (spanning ten days), and it carried a six-man crew—including Ulf Merbold, the first European to fly aboard the shuttle. Spacelab-1 showed just how far we have come in the last decade. It carried as much scientific equipment as Skylab did ten years earlier, yet gathered fifty times as much information in a ten-day mission than Skylab did

Cutaway view of Spacelab's shirtsleeve environment.

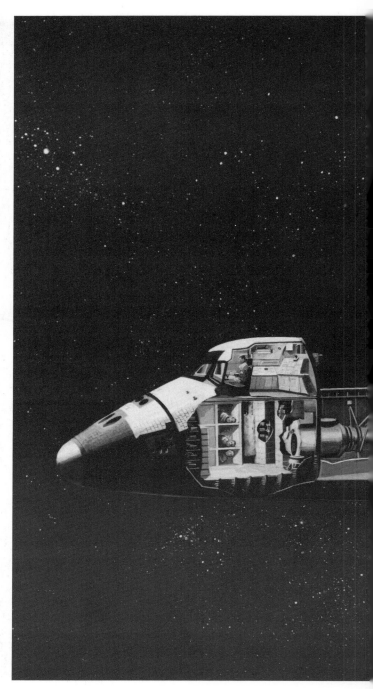

over 172 days of manned operation.

Future Spacelab missions will be more specialized, focusing each mission on related research areas. For instance, Spacelab-2 investigated astronomy and solar physics, while Spacelab-3 concentrated on materials processing and life sciences. In this way the shuttle will only have to be pointed in one direction (toward the earth or out into deep space) enabling better utilization of Spacelab experiments.

Ground-breaking results were realized during this mission on many experiments, especially in metallurgy and biology. Research on never-before-produced alloys, such as lead-zinc and zinc-aluminum, was conducted that may lead to lighter, stronger metals. In the biological area, protein crystals were grown to sizes one thousand times larger than earth-based processes. Crystals this large have wide applications in molecular biology for determining enzyme structure. By allowing scientists to understand enzyme structure, these large crystals play a crucial role in the development of new medicines. Following these crystal-growth breakthroughs, dozens of universities and pharmaceutical companies have proposed experiments that will fly on the shuttle.

Both the pressurized module and the pallets are reusable. Five shuttle-Spacelab missions using the pressurized module are planned through 1988 and thirty-six missions will make use of the Spacelab pallets.

Long Duration Exposure Facility

Another unique payload taking advantage of the zero-gravity environment is the long duration exposure facility (LDEF). The 30-foot long LDEF can hold up to eighty-six experiments and provides an easy and economical platform for conducting experiments that require extended exposure to the space environment.

The LDEF is placed into a 300-mile orbit by the shuttle for periods of six months or more depending on mission requirements. It is then retrieved during a later shuttle mission for return to earth, and, like the shuttle, is reusable. The first LDEF mission began in April 1984 and carried over fifty separate experiments. After spending almost two years in space, the experiments will be returned to earth for extensive analysis to determine the effects of long-term spaceflight.

NASA, with an eye toward building a permanently orbiting space station, is one of LDEF's most frequent users and is testing many different materials aboard LDEF to see how they react to long duration exposure to the space environment. Experiments have included solar cells, electronic parts, optical equipment, and various spacecraft materials that will be used to build advanced satellites as well as the space station.

Foreign Competition

Foreign activity in space manufacturing has been vigorous. The Soviets are the most active. Their space stations have given them an ideal environment to conduct experiments. A 1980 report by the General Accounting Office (GAO) of Congress estimated that the Soviets have some 350 top scientists actively engaged in space-related materials research. This was even before the launch of *Salyut 7* and its record-breaking long duration flights.

The European nations are also quite active in space manufacturing research and conducted many of the early Spacelab experiments. In fact, West Germany has such an abundance of experiments planned that it purchased two entire Spacelab flights for its own use, one in 1985 and another scheduled for late 1988. Japan has also invested extensively in materials research and has reserved a Spacelab mission of its own, scheduled for early 1988. The GAO report concluded that foreign

competition and possible dominance of space manufacturing are a real threat.

The problem is that the required investments are large and the returns are long-term and not always obvious. For instance, the McDonnell Douglas–Johnson & Johnson project began in 1977 and has required an investment of over $60 million. Given these figures industry has been understandably slow to react. Optimistically, the early successes aboard the shuttle will spur both government and

Long-duration exposure facility high above the Gulf of Mexico prior to release by the shuttle's robot arm.

industry to increase research funding for space manufacturing.

The examples of space manufacturing presented here only scratch the surface of future possibilities. Before this century is over space factories could be routinely producing materials and products that aren't even envisioned today.

CHAPTER 9

SCIENTIFIC EXPLORATION OF SPACE

"THROUGH THE EYES OF THE AMERICAN SPACECRAFT VOYAGER 1 PASSING NEAR THE GIANT PLANET SATURN, THE WORLD HAS LEARNED MORE ABOUT THAT PLANET IN ONE WEEK THAN IN ALL PREVIOUS RECORDED HISTORY."

SENATOR STROM THURMOND
NOVEMBER 17, 1980

Some of the most spectacular returns from the space program have come from scientific missions, particularly planetary probes. Who could forget the dramatic photos of Jupiter and Saturn that the Voyager spacecraft radioed back to earth in 1979, 1980, and 1981?

As breathtaking and beautiful as the photographs have been, if pictures were the only expected return, those missions would never have been undertaken. Planetary missions are stepping stones to try and determine the nature, origin, and evolution of our solar system and the universe. By looking at other planets and beyond we can begin to better understand the earth and the effect the

Jupiter and two of its moons, Io *(above)* and Europa *(below)*, as photographed by *Voyager 1,* March 1979.

universe has on our lives. For instance, the sun is the source of all energy on earth. How is this energy generated, transformed, and transported through space? How does it affect us and the rest of the solar system?

During the Skylab missions in 1973 and 1974, astronauts were able to observe the sun for extended periods of time without the distorting atmospheric effects that plague earth-based

Computer-enhanced color image of Saturn — *Voyager 1,* October 1980.

SOLAR FLARE

SIZE OF EARTH

Above: Spectacular solar flare reaching a height of 375,000 miles, as photographed by Skylab cameras.

Below: A 76-year journey of Halley's Comet.

telescopes. They observed and photographed enormous solar flares. These solar flares occur with some regularity and affect the earth's weather in ways not totally understood, causing havoc among farmers (as well as meteorologists). Continued study of the sun will attempt to identify patterns in the sun's activity to aid in under-

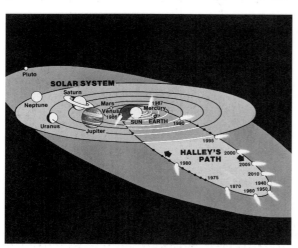

A Solar System Primer

Where does space begin?

Space is generally considered to begin at the upper edge of the atmosphere, approximately 80 miles above the earth's surface.

What is the solar system?

Our solar system is composed of the sun, which is a star, the nine planets and their moons, asteroids, meteroids, and comets. Our solar system is part of the Milky Way galaxy which is one of the billions of galaxies in the universe.

How big is the sun?

The distance around the sun's equator is approximately 2,725,000 miles, or 109 times as large as the earth's 25,000 mile circumference. The moon is approximately 6,800 miles in circumference.

What is an asteroid?

Asteroids are small celestial bodies, also known as minor planets. There are over 30,000 asteroids, most of which are located between the planets Mars and Jupiter in what is known as the asteroid belt. Their sizes range from a fraction of a mile to over 1,500 miles in circumference.

standing its effects on earth's climate. Spacecraft have tracked the solar wind, solar cosmic rays, and the solar magnetic field, all of which have been found to extend beyond the orbit of Jupiter. How do these phenomena affect the earth? The other planets?

Pieces of the puzzle are beginning to fall into place but the universe is constantly changing. Exploration becomes more and more compelling as each discovery prompts new questions. The shuttle will launch new planetary spacecraft and large telescopes that will give us a new perspective of our solar system and the universe. First, let's review some of the scientific strides the space program has made to date.

Pioneering Planetary Exploration

As the accompanying table illustrates, the past twenty years have been the golden age of planetary exploration. Many of the spacecraft listed continue to return valuable data as they hurtle through space. *Pioneer 10*—which was launched in 1972 to study Jupiter and the asteroid belt—has been one of the most remarkable. After accomplishing its primary objectives *Pioneer 10* became the first man-made object to leave the solar system as it passed the orbits of Neptune and Pluto and entered interstellar space on June 13, 1983. Having travelled over 3.5 *billion* miles to that point, it is still transmitting data on gravity waves and whatever else it might find beyond the outer regions of the solar system.

Originally built to last eighteen months, *Pioneer 10* is now expected to continue transmitting for another ten years. During that time it may encounter the oft-theorized tenth planet. Moreover, in the vacuum of space, *Pioneer 10* will theoretically last forever. If it doesn't have any collisions, this cosmic mailman will still be travelling billions

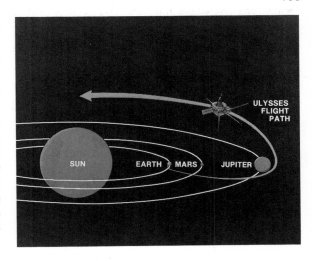

Above: Ulysses—the slingshot approach to the sun's poles. *Below: Galileo's* route to Jupiter.

of years from now when our own sun has burned itself out and civilization has left this planet.

Voyager 2 is another workhorse of the planetary program. Launched in 1977, its encounters with Jupiter and Saturn are well documented. It is now enroute to the outer planets where it will make close flyby studies of Uranus in 1986, and Neptune in 1989 before leaving the solar system.

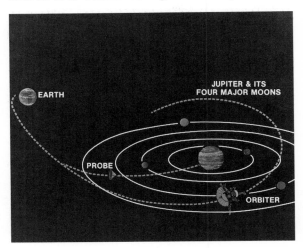

Opposite: Galileo probe will become the first manmade object to pierce Jupiter's atmosphere.
Above: Artist's concept of the Venus Radar Mapper and (right) the Mars Geoscience–Climatology Orbiter. *See page 111.*

The Shuttle and Scientific Exploration

The shuttle gives scientists a valuable new tool to further the study of near-earth phenomena as well as in the farthest reaches of the universe. Experiments aboard the shuttle will explore our universe as never before, studying regions of the

Planetary Missions

Spacecraft	Launch Date	Objective	Arrival Date	Closest Approach	Status and Accomplishments
Mariner 2	Aug 26, 1962	Venus flyby	Dec 14, 1962	20,900 miles	In solar orbit; first successful Venus flyby
Mariner 4	Nov 28, 1964	Mars flyby	Jul 14, 1965	6,118 miles	In solar orbit; returned 21 pictures of Mars
Mariner 5	Jun 14, 1967	Venus flyby	Nov 19, 1967	2,480 miles	In solar orbit; Venus flyby
Mariner 6	Feb 24, 1969	Mars flyby	Jul 31, 1969	2,120 miles	Heliocentric orbit; flew over equator of Mars, sent 75 TV pictures (see *Mariner 7*)
Mariner 7	Mar 27, 1969	Mars flyby	Aug 5, 1969	2,190 miles	Heliocentric orbit; flew over Martian southern hemisphere; sent 126 TV pictures, 33 of south polar region (see *Mariner 6*)
Mariner 9	May 30, 1971	Mars studies from orbit	Nov 13, 1971	1,025 miles	In Mars orbit; operated from 11-13-71 to 10-27-72; returned 7,329 pictures of Mars mapping 100% of the surface
Pioneer 10	Mar 10, 1972	Study Jupiter and the asteroid belt	Dec 4, 1973	82,000 miles	First spacecraft beyond Mars; first to transmit the astroid belt; first reconnaissance of Jupiter; first to cross orbit of Uranus; first to leave solar system
Pioneer 11	Apr 5, 1973	Study Jupiter and the asteroid belt	Dec 3, 1974	26,600	Second Jupiter probe; first to Saturn, 9-1-79

electromagnetic spectrum ranging from the infrared to ultraviolet, X-ray, and even gamma-ray phenomena. Once again it is the shuttle's ability to carry an experiment into space and return the hardware for reuse or upgrading that is proving to be a boon for researchers.

Three such shuttle missions will study the most legendary of all members of our solar system, Halley's Comet. Its return to earth in 1986, after a seventy-six-year absence, will be one of the most observed occurrences of the decade if not the century. As it approaches the earth, no less than five spacecraft—two Japanese, two Soviet, and one from the European Space Agency—will visit Halley's. The United States had planned to send its own Halley probe, but funding limitations resulted in its cancellation.

Fortunately, the shuttle enables NASA to undertake three relatively inexpensive observation experiments that will allow the United States to participate in this historic event. The three experiments for observing Halley's from aboard the shuttle are known as Astro-1, -2, and -3 (short for Astronomy Research Laboratory).

The Astro missions will use ultraviolet telescopes mounted on Spacelab pallets to measure

Spacecraft	Launch Date	Objective	Arrival Date	Closest Approach	Status and Accomplishments
Mariner 10	Nov 3, 1973	Venus & Mercury encounter	Feb 5, 1974 Mar 29, 1974	Venus 3,300 miles Mercury, 600 miles	6,800 pictures of Venus, Mercury, earth, and moon; first pictures of Venus; first mission to Mercury
Viking 1	Aug 20, 1975	Study Mars from its surface & orbit	Jul 20, 1976	Soft-landing & orbiting	Investigated Mars from orbit and surface; and search for life
Viking 2	Sep 9, 1975	Study Mars from its surface & orbit	Sep 3, 1976	Soft landing & orbiting	(See *Viking 1*)
Voyager 2	Aug 20, 1977	Jupiter-Saturn flyby	Jul 9, 1979	404,000 miles	Fourth Jupiter probe, 3rd to Saturn; will flyby Uranus Jan 1986, to study planet, its rings, and its moons; Neptune in 1989
Voyager 1	Sep 5, 1977	Jupiter-Saturn flyby	Mar 5, 1979	173,000 miles	Third Jupiter probe; 18,000 pictures of Jupiter and its moons; discovered ring around Jupiter and volcanic eruptions on Io; 2nd to Saturn, Nov 1980.
Pioneer-Venus I	May 20, 1978	Venus orbiter	Dec 4, 1978	29,000 miles	First U.S. detection of continuous lightning in atmosphere, confirming earlier Russian findings
Pioneer-Venus II Multiprobe	Aug 8, 1978	Venus lander	Dec 8, 1978	Descent to surface	Multiprobe separated into five instrumented probes that measured Venus atmosphere during descent to surface

Halley's thermal properties and ultraviolet wave lengths while photographing the comet. Comets are believed to be made of ice, rock, and gases that contain original matter from which the solar system was formed. These missions will attempt to pull back Halley's veil of secrecy and confirm the theory. The missions are planned for March and November 1986 and July 1987 as Halley's approaches, passes, and once again moves away from the earth. In addition, international cooperation and sharing of data will contribute to the understanding of this mystical celestial visitor which will not return again until the year 2061.

Shuttle Planetary Payloads

The shuttle will also launch planetary spacecraft. Two such missions already booked on the shuttle will turn Kennedy Space Center into one hectic place in May 1986 (or June 1987, depending on the outcome of the investigation of *Challenger's* accident). Because they both encounter Jupiter, the *Ulysses*–Solar-Polar mission and the *Galileo*-Jupiter mission must be launched with-

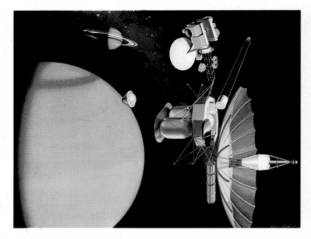

Above: Artist's concept of Titan Probe–Radar Mapper spacecraft.

Right: Shuttle astronauts servicing the Space Telescope.

in two weeks of each other to ensure optimum performance.

The *Ulysses*–Solar-Polar spacecraft is being built by the European Space Agency. Its primary objective is to explore the polar regions of the sun which have never been studied and cannot be seen from earth. To reach the sun's poles, the *Ulysses* spacecraft must use some clever orbital geometry. It must first travel to Jupiter to get a gravity assist or "slingshot" from the giant planet, boosting the spacecraft into a trajectory high enough to approach the sun's poles. The entire journey will take approximately four years with *Ulysses* scheduled to reach the sun in July 1990.

Instruments aboard *Ulysses* will measure the sun's radiation, particle flow, magnetic field, and surface activity. Scientists also hope to learn more about the processes of the sun that cause variations in our weather. Estimates show that a 2 percent change in the sun's energy output could have catastrophic effects on the earth. A 2 percent *increase* could cause accelerated melting of the earth's polar caps resulting in widespread flooding of coastal areas as the oceans rise. A 2 percent

decrease could create another ice age.

The *Galileo* spacecraft—named after the seventeenth-century astronomer who discovered and named Jupiter's four major moons—will be the first to enter the atmosphere of the giant shrouded planet Jupiter, 500 million miles from earth. *Galileo* will conduct a far more comprehensive investigation of Jupiter and its moons than was possible with the earlier Voyager missions, which were basically flybys. *Galileo* is actually a dual spacecraft mission; it has an *orbiter* and a *probe*.

Galileo is expected to encounter Jupiter in December 1988. Approximately 150 days before reaching Jupiter, the spacecraft will separate in two. The probe will take a trajectory sending it through Jupiter's atmosphere, while the orbiter will take a trajectory orbiting the giant planet.

A parachute will slow the probe's descent as it becomes the first man-made object to enter Jupiter's atmosphere. There the probe's seven instruments will measure the chemical composition and physical state of Jupiter's atmosphere and clouds. The probe is expected to transmit for approximately fifty minutes before being destroyed by its encounter with intense atmospheric pressure (up to twenty-five times greater than the earth's) and the ever increasing temperatures that will reach nearly 1,000°F.

Meanwhile, *Galileo's* orbiter will relay data from the probe to scientists on earth. Then for the next two years, the orbiter will carry out the first extensive study of Jupiter and its four major moons, Io, Europa, Ganymede, and Callisto. During this reconnaissance, *Galileo* will be able to fly one-hundred times closer to Jupiter's moons than the Voyager spacecraft. It will get within 172 miles of Callisto; 293 miles of Europa; 5l7 miles of Ganymede; and 620 miles of Io. The orbiter is expected to return over 50,000 high resolution photographs.

The 43-foot Space Telescope during final assembly and checkout.

Future Planetary Missions

To focus support for planetary missions, NASA formed the Solar System Exploration Committee (SSEC) in 1980. Composed of some of the greatest scientific minds in the country, the SSEC was called upon to develop a long-term approach to solar system exploration extending into the twenty-first century.

To increase the number of planetary missions, the SSEC recommended that future missions have more focused objectives. Earlier missions cost more because they tried to accomplish every objective in a single mission. By concentrating on one or two objectives per mission the SSEC believes that twelve to fifteen planetary missions could be conducted before the end of this century. The SSEC recommended a balanced approach to solar system exploration that would include missions to the inner planets, the small bodies (asteroids and comets), and the outer planets. Four missions have received top priority as initial goals in this new approach:

1. the Venus Radar Mapper
2. the Mars Geoscience-Climatology Orbiter
3. the Comet Rendezvous-Asteroid Flyby
4. the Titan Probe–Radar Mapper

Already the Venus and Mars missions have received approval and initial funding from Congress. The Venus mission, scheduled for shuttle launch in 1988, will use radar to penetrate Venus's gaseous atmosphere and map the hidden surface below. The Mars mission, to be launched in 1989, will focus on the planet's geoscience, measuring its elements and composition on a global scale. We already have optical maps of Mars from the early Mariner and Viking missions, but compositional mapping would provide new dimensions to our understanding of the Red Planet.

Space Telescope following shuttle deployment.

The Comet Rendezvous-Asteroid Flyby mission is designed to study comets and asteroids at close range. These small bodies are believed to consist of high-grade minerals that could someday provide raw materials for large-scale construction in space. Thousands of asteroids are located between Mars and Jupiter; they range in size from a fraction of a mile to 500 miles across.

The Titan mission would use radar to map Titan, the largest of Saturn's moons. Titan is the only body in the solar system with an atmosphere similar to that theorized to exist around the earth in its pre-life stage.

Four underlying goals support the SSEC's recommendations to conduct these missions. The primary goal continues to be the determination of the origin, evolution, and present state of the solar system. Although great progress has been made towards this goal in the last two decades, many questions remain. Two additional goals include understanding the earth through comparative planetary studies and understanding the relationship between the chemical and physical evolution of the solar system and the appearance of life, both of which require intensive study of other solar-system bodies. The fourth goal is to survey the minerals potential of asteroids and comets for use on earth, or for space construction.

The Space Telescope

Without doubt, one of the most promising of all shuttle payloads is the Hubble Space Telescope. It is named after Edwin Hubble, one of America's foremost astronomy pioneers, who in the 1920s proved the existence of other galaxies and is credited with determining the proper size of our own solar system.

Scheduled for launch in late 1986, the Hubble Space Telescope will be placed in an orbital altitude of 320 miles, high above the optically degrading effects of the earth's atmosphere. From that orbit, the 43-foot-long telescope will be able to see *seven times* farther and detect objects *fifty times* fainter than ground-based telescopes.

The telescope will open up an entirely new view of the universe. There is no limit to the realm of possible findings it may uncover. It may enable astronomers to see the very edge of the universe— nearly 14 billion light-years away. To try to put that immense number into perspective, consider our sun, which at 93 million miles from earth is only slightly less than 8½ *light-minutes* from us. (Since light travels at 186,000 miles per second, the distance travelled in a light-minute is 11,160,000 miles.) Or consider *Pioneer 10,* which had traveled 3.5 billion miles or only 5¼ *light-hours* when it left our solar system in June 1983.

In astronomy clarity of detail is just as important as distance. With the Hubble Space Telescope, scientists can study as never before the hundreds of relatively nearby stars (15,000 light-years or less) in search of solar systems similar to our own. Perhaps these instruments will discover enough about the basic physical processes in other solar systems to help us determine the likeliness of other life-supporting planets.

Furthermore, because it will be above the atmosphere and not limited by the day-night cycle, the telescope will operate nearly twenty-four hours a day. This will enable astronomers to observe long periods of cosmic activity instead of the occasional glimpses ground-based telescopes now provide.

The shuttle will not only launch the telescope into orbit, but shuttle astronauts will perform routine maintenance, replace instruments with updated versions, and make repairs as necessary. Thus, the shuttle will extend the uses and the life of the telescope, largely insuring its projected fifteen-year minimum operational lifetime—into the twenty-first century.

The exploration of the universe from the vista of space has greatly expanded mankind's knowledge and imagination. Many of the mysteries of the universe—pulsars, quasars, exploding galaxies, and black holes—are beginning to be understood. The space shuttle's ability to place large payloads such as the Hubble Space Telescope above the atmosphere will forever change the way we look at the universe.

CHAPTER 10

SPACE: THE LIMITLESS OPPORTUNITY

"OUR CIVILIZATION IS LIKE A BABY IN THE WOMB, RUNNING OUT OF ROOM TO GROW—WHOSE SOURCE IS DIMINISHING AND IS FEELING AFRAID. WHAT IT DOESN'T KNOW IS THAT IT IS ABOUT TO BE BORN. AS A SPECIES WE MAY BE ON THE THRESHOLD OF OUR MATURITY— ABOUT TO MOVE OUT OF OUR MOTHER'S WOMB."

KRAFFT EHRICKE
(Pioneering German Rocket Scientist)

Once feared as a vast, lifeless void, space is now a base for worldwide communications, earth-observation satellites, materials processing, and a multitude of scientific endeavors. Benefits from the space program are influencing our world daily . . . and we have barely begun to tap the enormous potential that space offers. In the coming decades space travel will become increasingly commonplace as permanent orbiting space stations are launched and serviced by the space shuttle.

"A space station . . . within a decade."

Now that the shuttle is an established system, the way we operate in space will change rapidly. Even though the loss of *Challenger* has taught us that space flight will never be routine, the overall success of the shuttle has shifted the attention to activities performed in space, not just the trip there.

Communications, scientific exploration, and defense will continue to provide the most visible benefits from space during the coming decade. But with the increased usage of the shuttle, new products will become available that will expand the use of space to many new participants. Programs such as Spacelab, in conjunction with the shuttle, will lay the groundwork for the future use of space. Within a decade the addition of a permanent manned space station will provide a base for unlimited space projects.

Space Station: The Next Giant Step

NASA has long envisioned the permanent occupancy of space, but has never been fully supported in that goal. The Skylab program of 1973–1974

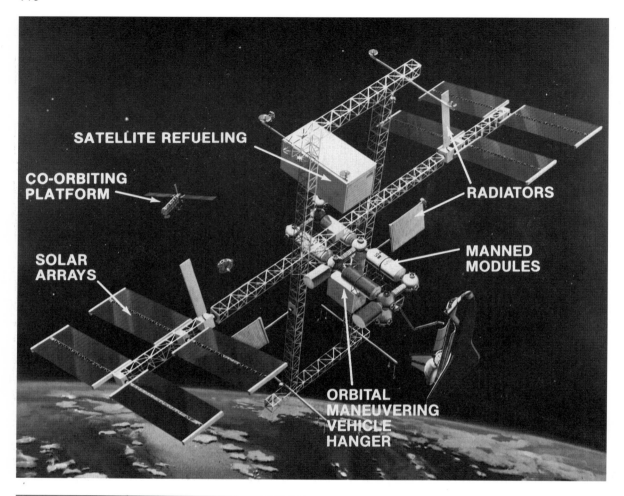

SATELLITE REFUELING

CO-ORBITING
PLATFORM

RADIATORS

SOLAR
ARRAYS

MANNED
MODULES

ORBITAL
MANEUVERING
VEHICLE
HANGER

Above: The initial space station and some of its basic components. This is NASA's reference configuration for the design phase of the program.

Opposite: A growth version of the station in the late 1990's with a full compliment of experiments.

was a good first step, but it was only a temporary effort. Today the reality of a permanently occupied space station is within reach. The ability to routinely "shuttle" to and from a space station, rotating crews and returning goods, has made the space station a near reality.

On January 25, 1984, during his State of the Union message, President Reagan directed NASA to begin developing a space station. In a speech reminiscent of John Kennedy's call for the Apollo program he said:

American has always been greatest when we dared to be great. We can reach for greatness again. We can follow our dreams to distant stars, living and working in space for peaceful, economic and scientific gain . . . tonight

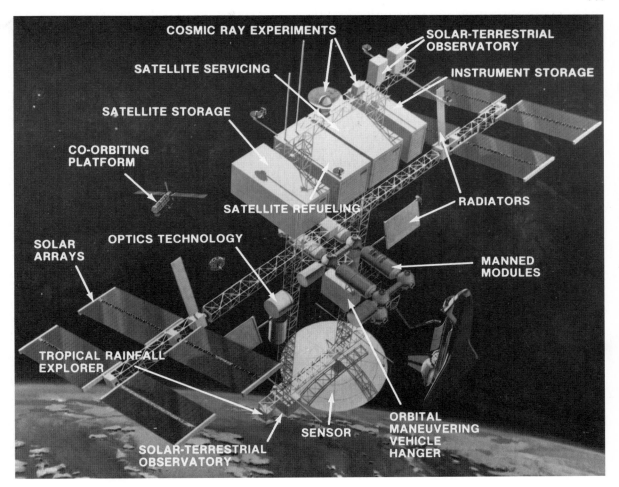

COSMIC RAY EXPERIMENTS

SOLAR-TERRESTRIAL OBSERVATORY

SATELLITE SERVICING

INSTRUMENT STORAGE

SATELLITE STORAGE

CO-ORBITING PLATFORM

RADIATORS

SATELLITE REFUELING

OPTICS TECHNOLOGY

SOLAR ARRAYS

MANNED MODULES

TROPICAL RAINFALL EXPLORER

SOLAR-TERRESTRIAL OBSERVATORY

SENSOR

ORBITAL MANEUVERING VEHICLE HANGER

I am directing NASA to develop a permanently manned space station— and to do it within a decade.

Space technology now calls for permanent facilities. Each step in the evolution of space has had an impact. The pioneering efforts of the Mercury and Gemini programs made Apollo possible. The enormous Saturn V that boosted Apollo enabled us to put something as large as Skylab into orbit to prove the feasibility of long-duration space operations. And the space shuttle has tied it all together.

It builds on all the advances made in the early programs and makes the space station possible.

Since the president's announcement, NASA has initiated detailed studies of a space station with ground fabrication expected to begin in 1987 leading to operation of the station between 1992 and 1994. Fittingly, 1992 coincides with the five hundredth anniversary of Columbus's discovery of the New World.

The shuttle, combined with a permanently manned space station, will support the future space activities of the United States and the free

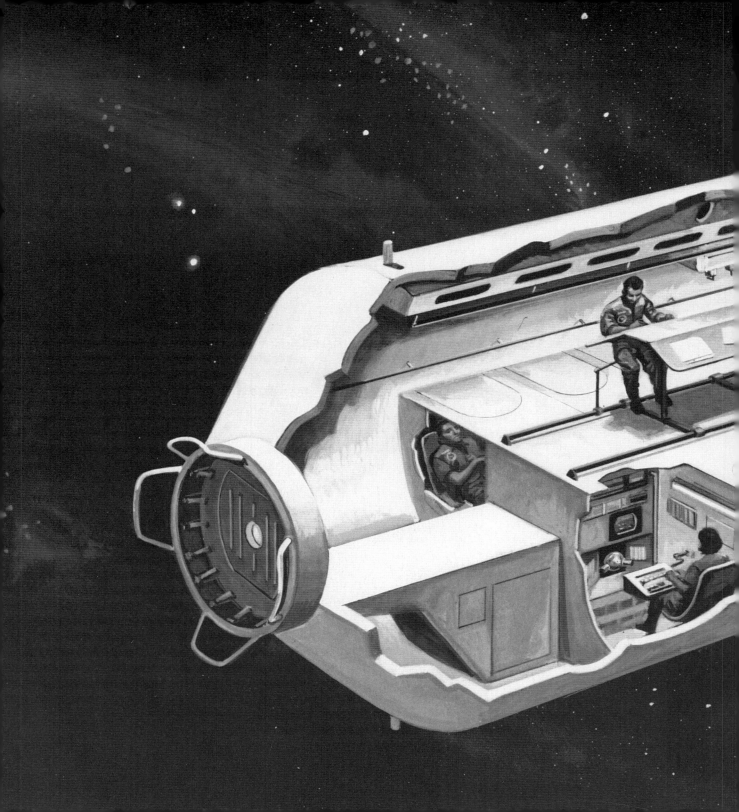

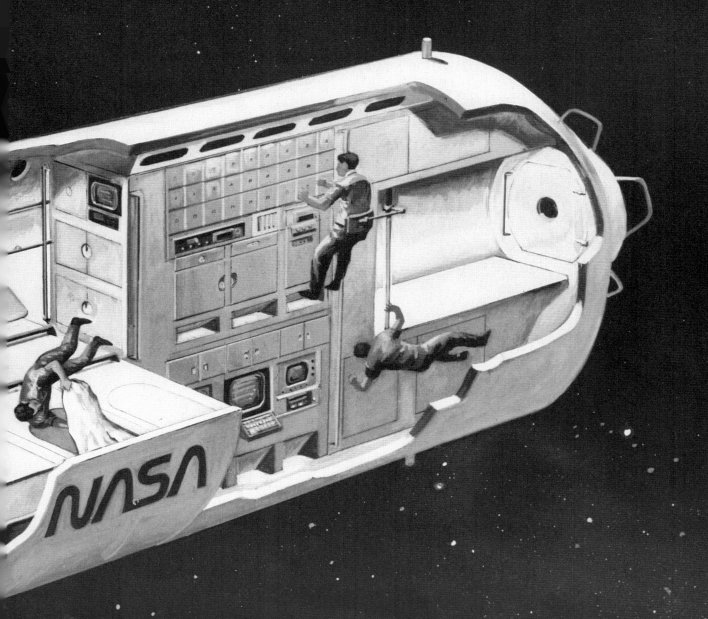

A typical habitability module of the space station.

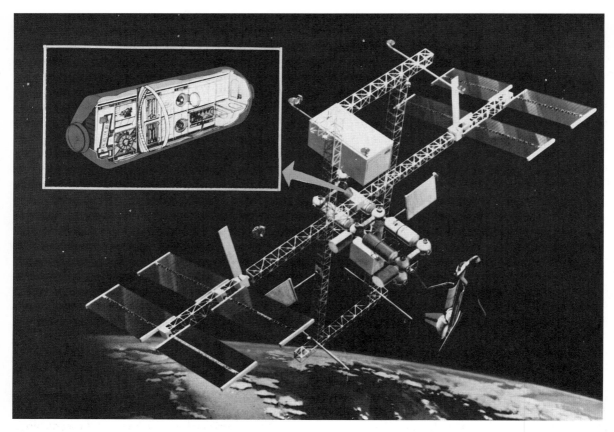

Typical space manufacturing module of the space station.

world. Already Europe, Canada, and Japan are studying the space station concept and have agreed to participate. A proposed space station of modular design would be assembled by shuttle astronauts. The modular design would allow for additional growth of the station as mission needs change through the rest of this century and into the next. As a scientific base, the space station will enable man to explore the cosmos as never before: to observe the sun, the planets, and the stars from the pollution-free environment of space on a continuous basis. For industry a space station will yield the opportunity to build on shuttle successes by taking advantage of the space environment for the creation of new products not available on earth. The station will also be used to assemble, service, and repair satellites. Finally, the station will serve as a staging point for later giant steps in space exploration such as a lunar base or manned exploration of Mars.

The initial station will provide living quarters and working facilities for a crew of eight astronauts; this could later be expanded to sixteen. The crews will be rotated every three to six months. Located at an altitude of 300 miles, the basic space station will consist of five 15-foot diameter modules that will be assembled in space during seven to nine shuttle missions. There will be two laboratory modules, one for materials processing, the other for life sciences; a logistics module will provide food, supplies, and maintenance equipment; and two habitability modules will provide the crew's living and working environment.

For electricity the initial station will require approximately 20,000 square feet of solar cells —two-thirds the size of a football field. And unlike Skylab, which went out of control and reentered the atmosphere in 1979, the new station will have its own propulsion system for orbit maintenance.

The Future

As the 1980s progress and spaceflight becomes routine there is a very real probability that space manufacturing will become a major space activity. Space processing of pharmaceuticals is already showing great promise and large quantities will be available beginning in the late 1980s. By the early to mid-1990s the space station will be producing quantities of products that are just now being contemplated.

The space station will provide the first step to large-scale space facilities that in the next century will call for thousands of space pioneers to live and work in space for extended periods of time. Much like the pioneers of the Old West, they will establish the initial settlements in space that will evolve into larger, more sophisticated facilities in the next century.

Interest in returning to the moon has grown significantly in recent years. Preliminary studies show that a lunar base could be in operation early in the next century (2005–2015). Such a base could serve as an international outpost and the first step in mankind's inevitable settlement of our solar system. As these settlements evolve into space cities, mining asteroids and the moon will become practical. While early space outposts will be prefabricated on earth and transported to space for assembly, later settlements will be made of extraterrestrial materials. The mere fact that it takes only one-sixth the thrust to escape lunar orbit than it does to escape earth orbit will make lunar mining economically attractive when construction in space or on the moon grows sufficiently.

Space projects, however, by their complex nature, require long development cycles and competitors could put the United States at a serious technological disadvantage if our efforts lag. Playing "catch-up ball" will be increasingly difficult. What is demanded now is the same sense of urgency and determination that characterized the Apollo program. In early space programs significant overlaps greatly accelerated the pace. Gemini was underway before Mercury was completed; Apollo was initiated while Gemini was still operative; and the shuttle was well into its planning phase while men were landing on the moon. The present pace lags by comparison. If the space program had continued the funding pace of Apollo there could have been a fleet of *twelve* shuttles and *two* space stations in operation *today*.

Space technology is developing rapidly and other countries, particularly the Soviet Union, have recognized the need to expand space activities, not only for their own benefit, but to influence world affairs. Space activities will contribute a growing percentage to the world economy. What's more, as developing countries realize the global benefits that space can provide, they will look for leadership to the country that supplies those benefits.

Today more than thirty years after Sputnik shocked the world and stirred the United States into action, it is clear that the space race is far from over. Expanded operations in space are inevitable—the choice is to lead or to follow. The words of President Kennedy come clearly to mind:

The exploration of space will go ahead whether we join it or not. It is one of the great adventures of our time and no nation that expects to be the leader of other nations can stay behind in the race for space.

Twenty-first century lunar base.

APPENDIX I

BECOMING A SHUTTLE ASTRONAUT

NASA's astronaut corps currently consists of over one hundred men and women. To meet the growing demand for space shuttle crews NASA initiated an annual astronaut selection beginning in 1983. There are two basic types of shuttle astronauts: *mission specialists,* who conduct experiments and-or launch satellites, and *pilot astronauts* who actually fly the shuttle. Each shuttle mission will carry a minimum of four astronauts and can carry as many as eight.

If you are chosen as an astronaut candidate, expect an initial one-year training program followed by one and a half to two years of advanced training before you are assigned to any specific flights. The training program covers three basic areas: technical, scientific, and shuttle systems.

Technical training covers a myriad of subjects including:

- Propulsion systems
- Flight performance
- Space navigation
- Avionic systems
- Communication and tracking systems

Scientific training includes seven disciplines of study:

- Astronomy
- Planetary science
- Earth observations
- Atmospheric science
- Space physics
- Materials science
- Medical science

Training on shuttle systems include:

- Overall orientation
- Orbiter systems
- Payload carriers
- Flight overview
- Command and control functions
- Logistic support
- Mission planning

In addition, astronauts are put through some very unique training as they are exposed to conditions simulating zero gravity before flight.

Astronauts Anna and Bill Fisher on a familiarization flight. The parabolic flight path of NASA's KC-135 aircraft provides up to five minutes of weightlessness.

NASA's two methods for imitating zero gravity are flying airplanes in parabolic flight trajectories, and scuba diving in special tanks at Johnson Space Center in Houston, Texas or at the Marshall Space Flight Center in Huntsville, Alabama.

NASA's minimum qualifications for *mission specialist* are: bachelor's degree in engineering, biology, mathematics, or a physical science, supplemented by at least three years in related professional experience, an advanced degree is desirable and may be substituted for the experience; ability to pass a NASA Class 2 space-flight physical; and height between 5' and 6'4".

Minimum requirements for *pilot astronaut* are: bachelor's degree in engineering, biology, mathematics, or a physical science; at least 1,000 hours of flight time as a pilot in command of high-performance jet aircraft with flight-test experience highly desirable; ability to pass a NASA Class 1 space-flight physical examination, similar to military and civilian flight physicals; and height between 5'4" and 6'4".

Qualified individuals interested in applying for shuttle positions as pilots or mission specialists can obtain forms and supplementary materials by writing to Astronaut (Mission Specialist) Candidate Program or Astronaut (Pilot) Candidate Program, NASA Johnson Space Center, Code: AHX, Astronaut Selection Office, Houston, Texas, 77058, U.S.A. Applications are accepted continuously, and selections are made in the spring.

APPENDIX II

THE NATIONAL AERONAUTICS AND SPACE ADMINISTRATION —RESEARCH, DEVELOPMENT, AND TEST FACILITIES

Most of NASA's facilities have visitor centers open to the public. Phone numbers are listed.

Ames Research Center, Moffett Field, California —Ames conducts laboratory and flight research such as atmospheric reentry, fundamental physics, solar physics, and planetary environments.

Hugh L. Dryden Flight Research Center, Edwards Air Force Base, California—Dryden's research covers manned flight within and outside the atmosphere, including low-speed supersonic, hypersonic, and reentry flight and aircraft operations. The approach and landing tests of the space shuttle orbiter were held here and many shuttle mis-

sions land here. (805) 258-3460. Museum, gift shop; open Monday-Friday.

Goddard Space Flight Center, Greenbelt, Maryland—Goddard is responsible for a broad variety of unmanned earth-orbiting satellites and sounding-rocket projects. Among its projects are Solar Max, the Explorer program, weather satellites, and Landsat. (301) 344-8981. Museum, gift shop; open Wednesday-Sunday.

Jet Propulsion Laboratory, Pasadena, California —JPL is operated for NASA under contract by the California Institute of Technology. The laboratory's primary role is investigation of the planets.

It has managed all the nation's planetary programs and manages the Voyager and Galileo programs.

Lyndon B. Johnson Space Center, Houston, Texas —JSC has been responsible for development of manned spacecraft since the inception of the space program and selects and trains astronauts. Mission control for manned spaceflight is also located at the center. (713) 483-4241. Museum, gift shop, open daily. Mission Control Center briefing every hour (except during missions).

John F. Kennedy Space Center, Florida—KSC prepares and launches manned and unmanned space vehicles for NASA. It has been the primary launch facility since the space program's inception and also serves as the shuttle's primary landing site. (305) 452-2121. Museum, gift shop, Imax theater; open daily. Bus tours (limited during launches).

Langley Research Center, Hampton, Virginia —Langley provides technology for manned and unmanned exploration of space and for improvement and extension of performance, utility, and safety of transport and military aircraft. Langley devotes more than half its efforts to aeronautics. (804) 865-2855. Museum, gift shop, theater; open daily.

Lewis Research Center, Cleveland, Ohio— Aircraft and rocket propulsion and energy systems for space and on earth are among Lewis's major programs. It is the main NASA center engaged in energy activities for the Department of Energy. (216) 267-1187. Museum, gift shop, theater; open daily.

George C. Marshall Space Flight Center, Huntsville, Alabama—Marshall serves as one of NASA's primary centers for the design and development of space transportation systems, orbital systems, scientific payloads, and other means for space exploration. It is here that Apollo's Saturn V rockets were developed. The center has major responsibilities for space shuttle development testing and fabrication, including the main engines and solid rocket boosters. Out of state 1-800-633-7280, in Alabama 800-572-7234. Museum, gift shop, spacedome theater; open daily.

NASA Headquarters, Washington, D.C.—NASA headquarters has overall responsibility for planning, coordination, and control of all the agency's activities. Headquarters is divided into six major offices headed by associate administrators who report directly to the NASA administrator. Each of the NASA field centers reports to one of the six offices. There are associate administrators for the following areas:

- Space flight
- Space station
- Space science and applications
- Astronautics and space technology
- Space tracking and data systems
- Management

National Space Technology Laboratories, Bay St. Louis, Mississippi—This complex conducts developmental tests of the space shuttle main engines. Hundreds of test firings of SSMEs have been held here during the shuttle's development. Today, tests continue on advanced engines. (601) 688-2370. Museum, gift shop, theater; open daily. Van tours.

Wallops Flight Center, Wallops Island, Virginia —Wallops is one of the oldest and busiest ranges in the world. Some 300 experiments are sent aloft each year on vehicles that vary in size from small sounding rockets to the four-stage Scout with orbital capability. (804) 824-3411, x298. Museum, gift shop, theater; open Thursday-Monday.

APPENDIX III

THE INDUSTRIAL TEAM THAT BUILT THE SHUTTLE

It takes a nationwide effort with over 3,000 contractors involved to build a shuttle. The three major companies responsible for constructing the shuttle are: *Rockwell International,* the prime contractor, which has two divisions contributing to the shuttle: Rockwell's Space Transportation Systems Division, Downey, California, builds the orbiter; Rockwell's Rocketdyne Division, Canoga Park, California, builds the main engines. *Morton-Thiokol Corporation,* Wastach Division of Brigham City, Utah—Solid rocket boosters. *Martin Marietta* of Denver at their New Orleans, Louisiana facility—external tank.

Other companies include:

Aerojet General, Aerojet Liquid Rocket Co., Sacramento, California
Orbital maneuvering subsystem engines

Airesearch Manufacturing Co., Garrett Corp., Torrance, California
Air data transducer assembly, computer, cabin air pressure safety valve, air shutoff solenoid valve, ground coolant unit

Autonetics Group, Rockwell International, Anaheim, California
Shuttle avionics test set

Avco Corp., Wilmington, Maine
Ku band antenna, Ku waveguide assembly, crew module bulkheads

Ball Brothers Research Corp., Boulder, Colorado
Star tracker, active keel activator, payload retention latch

B.F. Goodrich Co., Troy, Ohio
Main nose landing gear, wheel main landing gear brake assembly, main and nose gear tires

Boeing Aerospace Co., Seattle, Washington
Sneak circuit analysis, carrier aircraft modification (747), load measurement system

Collins Avionics Group, Rockwell International, Cedar Rapids, Iowa
Display driver unit, horizon situation indicator

Columbus Aircraft Division, Rockwell International, Columbus, Ohio
    Body flap structure, nose gear doors

Corning Glass, Corning, New York
    Windshield, windows

Fairchild Republic, Farmingdale, New York
    Vertical tail

General Dynamics, Convair Aerospace Division, San Diego, California
    Mid fuselage

Grumman Corp., Bethpage, New York
    Wings, launch processing system

Honeywell, Inc., St. Petersburg, Florida
    Flight control subsystem, rotational hand controller, speed brake thrust control, translation hand control, reaction jets, aerosurface servo amplifiers

IBM Corp., Federal Systems Division, Oswego, New York
    General purpose computer, CRT display subsystem, input-output processor

Johns Manville Co., Mansville, New Jersey
    Thermal protection system reusable blankets

Morton-Thiokol, Wastach Division, Brigham City, Utah
    Solid rocket motors

Lockheed-California Co., Burbank, California
    Ejection seats, orbiter structure testing

Lockheed-Missiles & Space Co., Inc., Sunnyvale, California
    Reusable surface insulation (tiles)

Los Angeles Division, Rockwell International, Los Angeles, California
    Aft fuselage thrust structure, crew module panels, ground vibration test model

Marquardt Company, Van Nuys, California
    Reaction control thrusters

Martin Marietta, Denver, Colorado
    External tank, manned maneuvering unit

McDonnell Douglas Astronautics Co., St. Louis, Missouri
    Orbital maneuvering subsystem-reaction control subsystem aft pod, cargo integration

Parker Hannifin Corp., Irvine, California
    Disconnects, valves, accumulators, couplings, valve pressure relief system

RCA Corp., Princeton, New Jersey
    Closed circuit television (orbiter and remote manipulator system)

Rocketdyne Division, Rockwell International, Canoga Park, California
    Space shuttle main engines

Singer Kearfott, Little Falls, New Jersey
    Inertial measurement unit, multiplexer interface adapter, data bus coupler

Space Transportation Systems Division, Rockwell International, Downey, California
    Space shuttle orbiter final assembly, overall systems integration

Spar Aerospace, Ltd., Toronto, Canada
    Remote manipulator system (robot arm)

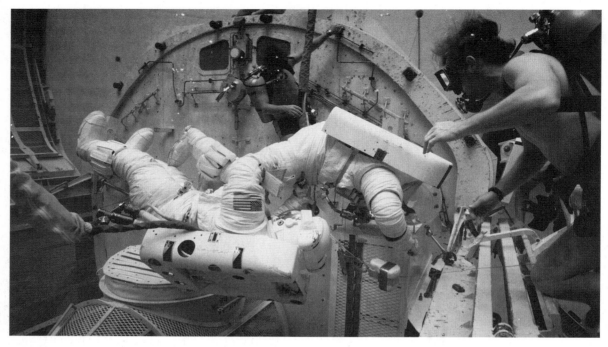

Astronauts John Young and Owen Garriott during training in the underwater weightless environment training facility of Johnson Space Center.

Sperry Rand Corp., Flight Systems Division, Phoenix, Arizona
Multiplexer-demultiplexer, automatic landing system, airspeed and altimeter indicators

Sundstrand Corp., Rockford, Illinois
Auxiliary power unit, actuation units, hydrogen recirculation pump assembly

TRW Electronics Systems Division, Redondo Beach, California
Tracking and data relay satellite system

Tulsa Division, Rockwell International, Tulsa, Oklahoma
Cargo bay doors, ground support equipment

United Space Boosters, Inc., Sunnyvale, California
Solid rocket booster recovery and refurbishment

United Technologies Corp., Power Systems Division, South Windsor, Connecticut
Fuel cell power plant

United Technologies Corp., Hamilton Standard Division, Windsor Locks, Connecticut
Atmospheric revitalization subsystem, flash evaporator system, extravehicular activity support system, mobility unit suits

Vought Corp., Dallas, Texas
Leading edge structure, nose cap, radiator, flow control assembly bulkhead

Opposite: STS-1 crew, Bob Crippen and John Young, during training.

Glossary

ALT   Approach and Landing Test: the series of unpowered flight tests that demonstrated the shuttle orbiter's basic air worthiness and handling characteristics (August–October, 1977).

ASTP   Apollo-Soyuz Test Project: joint U.S.-Soviet mission that enabled rendezvous, docking, and cooperative zero-gravity experiments. The first international manned mission (July 1975).

EAFB   Edwards Air Force Base: located 100 miles north of Los Angeles, the shuttle's West Coast landing site and home of NASA's Dryden Flight Research Facility.

ESA   European Space Agency: a consortium of twelve European nations who have worked together on numerous space projects. Spacelab is one of ESA's major programs.

ET   External Tank: the large center tank attached to the orbiter, which supplies over 500,000 gallons of liquid hydrogen and oxygen to the SSMEs.

EVA   Extravehicular Activity: any astronaut activity outside the confines of a pressurized spacecraft.

GPS   Global Positioning System: a constellation of satellites—18 when fully operational in 1988—that provides unprecedented navigation accuracy. Also known as Navstar.

LDEF   Long Duration Exposure Facility: a large shuttle-deployed satellite (30 ft × 14 ft) that can carry up to 86 experiments requiring long-term exposure (up to 2 years) to the space environment.

MMU   Manned Maneuvering Unit: the backpack used by shuttle astronauts to conduct EVAs beyond the immediate vicinity of the orbiter's cargo bay.

OMS   Orbital Maneuvering System: the two engines located above the SSMEs which provide the required thrust to make orbital changes and to slow the orbiter down for reentry.

RCS   Reaction Control System: a series of 44 small thrusters used to make small orbital maneuvers and changes to the orbiter's pointing direction. A typical RCS engine provides 870 lbs of thrust compared to an OMS engine's 6,000 lbs.

SPAS   Shuttle Pallet Satellite: a reusable German-built satellite that can carry several different experiments. It flew on shuttle missions STS-7 and STS-11.

SRB   Solid Rocket Boosters: the two large boosters attached to the ET. Each SRB provides 2,900,000 lbs of thrust during the two minutes of burn time; they are then jettisoned for retrieval by ship for refurbishment and reuse.

SSME   Space Shuttle Main Engines: the three large engines located at the rear of the orbiter. Each SSME provides 375,000 lbs of thrust. They burn for the entire 8½ minutes of the launch phase and are shut off just prior to entering orbit. SSMEs are designed for reuse up to 55 times.

STS   Space Transportation System: the formal name for the space shuttle and all its components such as Spacelab, LDEF, and upper stages.

TDRS   Tracking and Data Relay Satellite: a series of six enormous satellites (57 ft × 16 ft when fully deployed) that will be launched by the shuttle during the 1980s. A network of three TDRS satellites can simultaneously connect up to 32 satellites and the shuttle with their ground stations.

TPS   Thermal Protection System: the tiles that protect the orbiter from the thermal extremes encountered during space flight. Reentry temperatures reach 2,400°F.

VAFB   Vandenberg Air Force Base: the shuttle's West Coast launch facility, located 120 miles northwest of Los Angeles. Its location permits shuttle launches into polar orbit.

ABOUT THE AUTHOR

George Torres, a native of New York City, is a long-time space enthusiast. He has had broad exposure to the space program and has written many articles on it. *A Quantum Leap* is his first book-length effort. He is currently employed at Rockwell International, prime contractor for the space shuttle and home of the Apollo command and service modules. Prior to joining Rockwell in 1979 Torres served in the U.S. Army, primarily at the 4th Military Intelligence Company, Colorado Springs. He is a member of the National Space Institute, the Planetary Society, and the American Institute of Aeronautics and Astronautics. He also holds bachelor's and master's degrees from California State University, Long Beach. Torres resides in Lakewood, California.